Studies in Computational Intelligence

Volume 1223

The series "Studies in Computational Intelligence" (SCI) publishes new developments and advances in the various areas of computational intelligence—quickly and with a high quality. The intent is to cover the theory, applications, and design methods of computational intelligence, as embedded in the fields of engineering, computer science, physics and life sciences, as well as the methodologies behind them. The series contains monographs, lecture notes and edited volumes in computational intelligence spanning the areas of neural networks, connectionist systems, genetic algorithms, evolutionary computation, artificial intelligence, cellular automata, self-organizing systems, soft computing, fuzzy systems, and hybrid intelligent systems. Of particular value to both the contributors and the readership are the short publication timeframe and the world-wide distribution, which enable both wide and rapid dissemination of research output.

Indexed by SCOPUS, DBLP, WTI AG (Switzerland), zbMATH, SCImago.

All books published in the series are submitted for consideration in Web of Science.

Oscar Castillo · Patricia Melin
Editors

Intelligent System Design Based on Soft Computing Models

Springer

Editors
Oscar Castillo
Division of Graduate Studies and Research
Tijuana Institute of Technology
Tijuana, Baja California, Mexico

Patricia Melin
Division of Graduate Studies and Research
Tijuana Institute of Technology
Tijuana, Baja California, Mexico

ISSN 1860-949X ISSN 1860-9503 (electronic)
Studies in Computational Intelligence
ISBN 978-3-031-97308-6 ISBN 978-3-031-97309-3 (eBook)
https://doi.org/10.1007/978-3-031-97309-3

This Springer imprint is published by the registered company Springer Nature Switzerland AG
The registered company address is: Gewerbestrasse 11, 6330 Cham, Switzerland

Preface

We describe, in this book, new directions on intelligent system design based on soft computing models, such as fuzzy logic, neural networks, and optimization algorithms, as well as their hybrid combinations. We also outline the application of the intelligent systems in areas such as intelligent control and robotics, pattern recognition, medical diagnosis, time series prediction, and optimization of complex problems. There are papers with the main topics from type-1 to type-3 fuzzy logic, which basically consists of a group of papers that propose new concepts and algorithms based on type-1, type-2, and type-3 fuzzy logic and their applications. There are also papers that present theory and practice of meta-heuristics in diverse application areas. There are interesting papers on different applications of fuzzy logic, neural networks, and hybrid intelligent systems in medical problems. In addition, we can find papers describing applications of fuzzy logic, neural networks, and meta-heuristics in robotics problems. Another set of papers are presenting theory and practice of neural networks in diverse application areas, including convolutional and deep learning neural networks. Finally, there are also a group of papers that present theory and practice of optimization and evolutionary algorithms in different areas of application.

In conclusion, the edited book comprises contributions on diverse aspects of fuzzy logic, neural networks, and optimization meta-heuristic algorithms for designing optimal intelligent systems and their application in areas such as, such as intelligent control and robotics, pattern recognition, decision-making time series prediction, and optimization of complex problems. We are hopeful that the book will serve as a guide for researchers and graduate students working in the computational intelligence area.

Tijuana, Mexico

Oscar Castillo
Patricia Melin

About This Book

In this book, new directions on intelligent system design based on soft computing models, such as fuzzy logic, neural networks, and optimization algorithms, as well as their hybrid combinations, are offered. In addition, the above-mentioned methods are discussed in application areas such as control and robotics, pattern recognition, medical diagnosis, decision-making, time series prediction, and optimization of complex problems. Nowadays, the main topics of this book are highly relevant, as most of the intelligent systems and hardware implementations in use today, manifest some form of intelligent feature to enhance their performance. In addition, theoretical speaking, new and advanced models, and algorithms of type-2 and type-3 fuzzy logic are presented, which will be in the near future of great interest to researchers in these areas. Also, novel nature-inspired optimization algorithms and innovative neural models are put forward in the manuscript, that are very popular subjects, at this moment. There are contributions on theoretical aspects as well as applications, which make the book very appealing to a wide audience, ranging from researchers, to professors and graduate students.

Contents

Fuzzy Logic

Sugeno-Based Fuzzy Systems for Accurate Heart Rate Level Classification

Ivette Miramontes and Patricia Melin

Abstract This paper presents the development and evaluation of a Sugeno Type-1 Fuzzy System for Heart Rate classification, integrating patient age and heart rate level as key inputs. Utilizing both trapezoidal and Gaussian membership functions, the study explores the effectiveness of these approaches in classifying risk levels. Experimental results from a dataset of 30 patients demonstrate that the trapezoidal model achieved a 100% accuracy rate, while the Gaussian model attained 93.33% accuracy, highlighting the potential for improvements in the latter due to instances of unclassified patients. The findings underscore the advantages of fuzzy systems in medical applications, emphasizing the need for refining membership functions and exploring Type-2 fuzzy systems to enhance performance in managing uncertainties. This work lays the groundwork for future research to expand the system's applicability and robustness in real-time clinical settings.

Keywords Fuzzy systems · Sugeno · Blood pressure · Heart rate · Classification

1 Introduction

The use of fuzzy systems in the field of medicine presents a significant opportunity to enhance the accuracy and speed of diagnoses, especially when integrating numerical data and clinical images [1–10]. One of the main advantages of fuzzy systems, such as the Sugeno Type-1 model, lies in their ability to simplify implementation in embedded devices [11–13]. This is particularly relevant for applications requiring real-time responses, such as continuous health monitoring of patients, where processing speed is crucial. In this work, we have designed a Sugeno Type-1 Fuzzy System for heart rate classification, aiming to analyze its performance and potential for future implementation in embedded systems. This approach not only optimizes response time but also maximizes resource efficiency, which is essential for

I. Miramontes (✉) · P. Melin
Tijuana Institute of Technology, Tijuana, BC, México
e-mail: cynthia.miramontes@tectijuana.edu.mx

O. Castillo and P. Melin (eds.), *Intelligent System Design Based on Soft Computing Models*, Studies in Computational Intelligence 1223,
https://doi.org/10.1007/978-3-031-97309-3_1

portable and low-power devices. Thus, we seek to demonstrate how the Sugeno model offers superior performance compared to other methods and explore its advantages in real-time medical applications.

The structure of this paper is outlined as follows: Sect. 2 provides a review of relevant literature, Sect. 3 details the methodology used, Sect. 4 presents the findings, and Sect. 5 concludes with the insights gained from the conducted tests.

2 Literature Review

This section presents the basic concepts necessary to understand the proposed method.

2.1 Blood Pressure and Hypertension

Blood pressure refers to the force exerted by circulating blood against the walls of blood vessels. This measurement is crucial because it facilitates the transportation of essential nutrients and oxygen throughout the body. Blood pressure consists of two key components: systolic pressure, which occurs when the heart contracts, and diastolic pressure, which takes place when the heart is at rest between beats. Blood pressure is expressed in millimeters of mercury (mmHg) [14–17].

Hypertension, commonly known as high blood pressure, refers to the chronic condition where blood pressure levels consistently exceed normal ranges. According to European guidelines for hypertension management, this is defined as systolic pressure exceeding 140 mmHg or diastolic pressure surpassing 90 mmHg [18–22].

2.2 Heart Rate

According to the World Health Organization (WHO), a normal resting heart rate for adults ranges from 60 to 100 beats per minute. Various factors can affect heart rate, including age, overall health, and lifestyle choices. For instance, habits such as engaging in high-intensity exercise, smoking, and alcohol consumption can play a significant role [23].

A heart rate below 60 beats per minute isn't always a cause for concern, especially if the individual is on certain medications like beta-blockers. Highly trained athletes may even have heart rates as low as 40 beats per minute due to their well-conditioned hearts, which do not need to work as hard to maintain a steady rhythm. However, for those who engage in moderate or low levels of exercise, heart rate changes are usually minimal.

An excessively fast or slow resting heart rate may signal serious medical issues, specifically tachycardia or bradycardia. Tachycardia, characterized by a resting heart rate above 100 beats per minute, can be linked to various conditions such as poor circulation, congenital heart defects, hypertension, anemia, and heart damage from events like heart attacks. Conversely, bradycardia, defined as a resting heart rate below 60 beats per minute in non-athletic individuals, may indicate heart damage that could lead to heart attacks, angina, low blood pressure, chronic inflammation, or endocarditis [24–28].

3 Methodology

A Sugeno Type-1 Fuzzy System has been designed for implementation within a neuro-fuzzy model [29–32], aimed at providing a diagnosis regarding an individual's risk of developing hypertension. This fuzzy system features two main inputs: the patient's age and heart rate. The "age" variable has a universe of discourse ranging from 0 to 100 years, segmented into four linguistic variables corresponding to different life stages: "Child," "Young," "Adult," and "Elder." On the other hand, the heart rate is defined within a range of 0 to 220 beats per minute and is divided into five membership functions representing different levels: "Very Low," "Low," "Normal," "High," and "Very High."

Figure 1 illustrates the trapezoidal membership functions used to model the inputs of the fuzzy system. The selection of these functions aims to effectively capture gradual changes in the variables and facilitate the classification of patients into different risk levels. This approach provides a clear and precise interpretation for assessing an individual's health status based on two key indicators—age and heart rate—allowing for early intervention if the risk of hypertension is elevated.

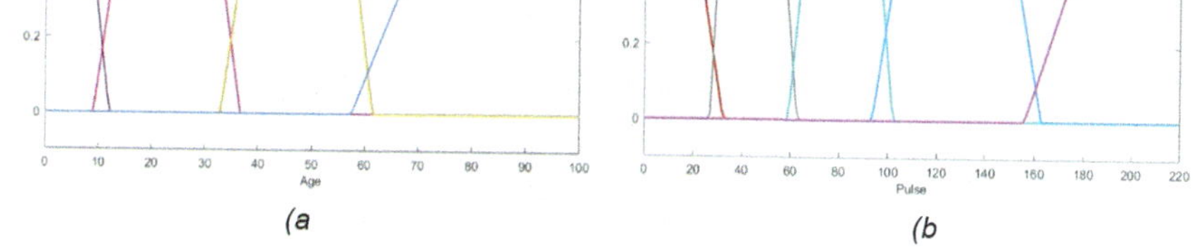

Fig. 1 **a** Input membership functions for age, **b** input membership functions for heart rate trend in the trapezoidal fuzzy system

The Table 1 presents the outputs obtained from the implemented fuzzy system. In this specific study case, a constant output has been chosen, which simplifies the interpretation of the results. The constant output value reflects a fixed decision rather than being subject to continuous variability, as could occur with more complex functions.

Figure 2 presents the output surface of the fuzzy model with trapezoidal membership functions. This three-dimensional graph illustrates how different combinations of the inputs "age" and "heart rate" affect the system's output. The generated surface provides a clear visualization of the transition areas between the various classification levels, enhancing the understanding of the model's behavior. This graphical approach is valuable for assessing the smoothness of transitions and the consistency of the system in classifying patients based on the defined parameters.

In a similar manner to the previously designed fuzzy system, a Sugeno-Type-1 Fuzzy model is implemented using Gaussian membership functions. Figure 3 shows

Table 1 Set of rules used in the Sugeno fuzzy system with trapezoidal membership functions

HR/age	Child	Young	Adult	Elder
Very low	5.1271	5.1271	5.1271	5.1271
Low	5.1271	20.4544	20.4544	35.2141
Normal	35.2141	35.2141	35.2141	35.2141
High	35.2141	56.6654	56.6654	89.2236
Very high	56.6654	89.2236	89.2236	89.2236

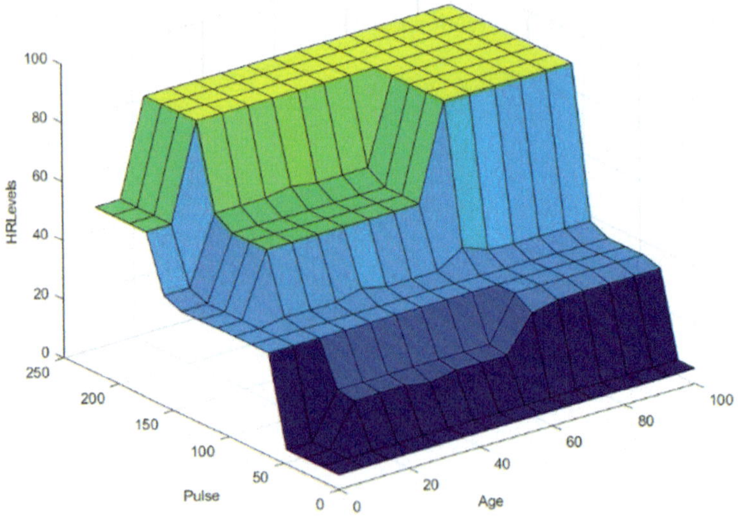

Fig. 2 Surface of trapezoidal fuzzy rules

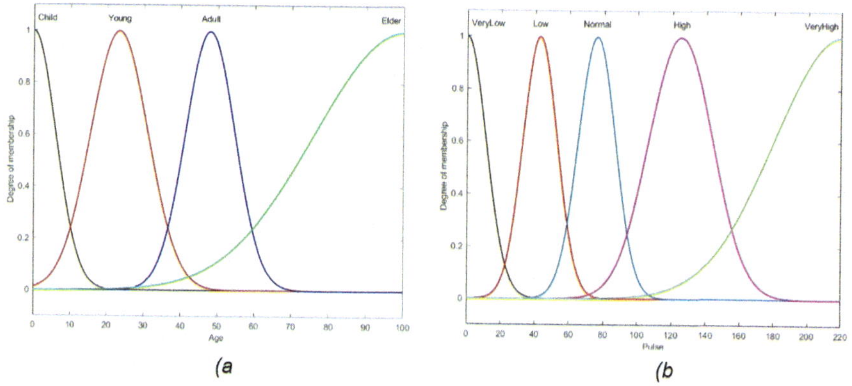

Fig. 3 **a** Input membership functions for age, **b** input membership functions for heart rate trend in the Gaussian fuzzy system

the inputs of this system, which are represented by smooth and continuous curves characteristic of Gaussian functions. These functions facilitate a more gradual transition between different input values, potentially resulting in more accurate classifications with reduced sensitivity to variations in the data. This approach is utilized to evaluate its performance in comparison to the model based on trapezoidal functions.

The output of the fuzzy model in question is also a constant, which simplifies its interpretation and analysis. These outputs are detailed in Table 2, providing a clear overview of the results obtained for each data set. Similar to other fuzzy approaches, using constant output facilitates decision-making based on predefined rules and ensures a stable and consistent response from the system to various inputs.

The surface of the fuzzy model with Gaussian membership functions is shown in Fig. 4. This figure allows for a visualization of the relationship between the inputs and the output of the system, providing a clear interpretation of the model's behavior in response to different combinations of age and heart rate.

Table 2 Set of rules used in the Sugeno fuzzy system with Gaussian membership functions

HR/age	Child	Young	Adult	Elder
Very low	4.442	4.442	4.442	4.442
Low	4.442	19.6513	19.6513	37.0519
Normal	37.0519	37.0519	37.0519	37.0519
High	37.0519	58.781	58.781	86.5992
Very high	58.781	86.5992	86.5992	86.5992

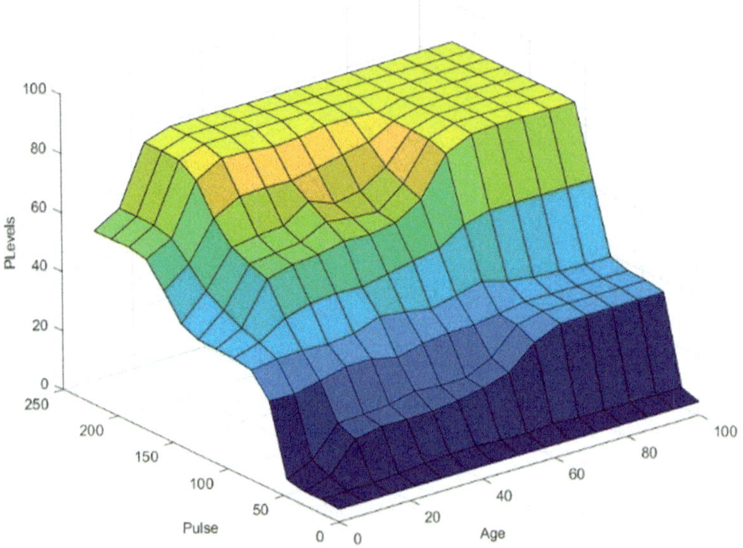

Fig. 4 Surface of Gaussian fuzzy rules

4 Results

To conduct the experimentation of the fuzzy systems and evaluate their performance, we selected a group of 30 patients from our database. For each patient, we collected their age and heart rate readings. These readings were processed through a neural network, which learned the behavioral patterns of the patients' heart rates and provided us with the observed trends. Subsequently, these trends were evaluated using the fuzzy inference system, generating the final classification results.

Table 3 presents the results of the fuzzy model utilizing trapezoidal membership functions. The fourth column displays the numerical results provided by the fuzzy system, while the fifth column presents the linguistic translation of those results. As shown, this model successfully classifies all patients, demonstrating the precision and effectiveness of the system in this specific context.

In Table 4, the results obtained from the fuzzy model using Gaussian membership functions are presented. The fourth column displays the numerical outputs generated by the fuzzy system, while the fifth column provides the linguistic interpretation of these results. Notably, two incorrect classifications were observed in this case, highlighted in bold for easy identification.

These types of errors emphasize the need to adjust or refine the parameters of the membership functions to minimize inconsistencies in future classifications. Despite these inaccuracies, the model remains robust, as the majority of predictions were correct, reinforcing its viability in classification applications like ours.

Table 3 Results obtained from the Sugeno fuzzy system with trapezoidal membership functions

No	Age	Heart rate	Output	Translation
1	82	203	89.224	Very high
2	12	205	89.224	Very high
3	63	57	35.214	Excellent
4	28	138	56.665	Above normal
5	96	214	89.224	Very high
6	15	215	89.224	Very high
7	96	127	89.224	Very high
8	80	65	35.214	Excellent
9	42	205	89.224	Very high
10	80	213	89.224	Very high
11	66	46	35.214	Excellent
12	85	209	89.224	Very high
13	68	177	89.224	Very high
14	75	110	89.224	Very high
15	66	70	35.214	Excellent
16	71	45	35.214	Excellent
17	27	48	20.454	Below normal
18	9	189	56.665	Above normal
19	70	97	50.937	Above normal
20	95	46	35.214	Excellent
21	44	109	56.665	Above normal
22	77	183	89.224	Very high
23	18	128	56.665	Above normal
24	45	156	57.439	Above normal
25	71	176	89.224	Very high
26	27	163	89.224	Very high
27	66	69	35.214	Excellent
28	12	130	56.665	Above normal
29	96	101	79.222	Very high
30	59	80	35.214	Excellent

To conduct a detailed comparison of the results obtained from the fuzzy system, confusion matrices were presented to analyze the model's performance. Table 5 displays the matrix corresponding to the fuzzy system with trapezoidal membership functions, where all correctly classified results can be observed.

This confusion matrix is a key tool, as it not only allows for the visualization of accurate predictions but also helps identify potential errors or areas for improvement

Table 4 Results obtained from the Sugeno fuzzy system with Gaussian membership functions

No	Age	Heart rate	Output	Translation
1	82	203	86.599	Very high
2	12	205	78.827	Very high
3	63	57	35.073	Excellent
4	28	138	63.236	Above normal
5	96	214	86.599	Very high
6	15	215	84.595	Very high
7	96	127	86.599	Very high
8	80	65	37.478	Excellent
9	42	205	86.591	Very high
10	80	213	86.599	Very high
11	66	46	35.917	Excellent
12	85	209	86.599	Very high
13	68	177	85.838	Very high
14	75	110	85.965	Very high
15	66	70	38.089	Excellent
16	71	45	36.937	Excellent
17	27	48	20.834	Below normal
18	9	189	68.727	Above normal
19	**70**	**97**	**70.477**	**Unrecognized value**
20	95	46	37.055	Excellent
21	44	109	62.083	Above normal
22	77	183	86.595	Very high
23	18	128	60.853	Above normal
24	**45**	**156**	**75.732**	**Unrecognized value**
25	71	176	86.434	Very high
26	27	163	79.245	Very high
27	66	69	37.747	Excellent
28	12	130	57.656	Above normal
29	96	101	79.822	Very high
30	59	80	43.547	Excellent

in the model. In this specific case, the model achieved 100% accuracy, correctly classifying all instances. This demonstrates that the trapezoidal membership functions used in this system are effective for classification tasks under the conditions of the experiment.

Furthermore, the clarity of the obtained results reinforces the model's robustness and validates the use of this type of membership function in applications requiring a high level of precision in classification. However, it is important to highlight that

Table 5 Confusion matrix for the Sugeno fuzzy system with trapezoidal membership functions

	Low	Below AV	Excellent	Above AV	Very high
Real low	0	0	0	0	0
Real below AV	0	1	0	0	0
Real excellent	0	0	8	0	0
Real above AV	0	0	0	7	0
Real very high	0	0	0	0	14

Table 6 Confusion matrix for the Sugeno fuzzy system with Gaussian membership functions

	Low	Below AV	Excellent	Above AV	Very high
Real low	0	0	0	0	0
Real below AV	0	1	0	0	0
Real excellent	0	0	8	0	0
Real above AV	0	0	0	5	0
Real very high	0	0	0	0	14

while this outcome is favorable, further testing with different datasets should be conducted to ensure the model's generalization across various clinical scenarios.

In Table 6, the confusion matrix corresponding to the fuzzy model with Gaussian membership functions is presented. Unlike other models, in this case, we can see that only 28 out of the 30 patients in our database were correctly classified. This indicates that two patients could not be categorized into any of the classes defined by the fuzzy system.

This classification omission may be attributed to various factors, such as the overlap of membership function ranges or insufficient information in the input data, leading to a lack of clear classification. It is important to note that such situations highlight one of the potential limitations of Gaussian functions; while they effectively capture the smoothness and continuity of variable changes, they can struggle with the accurate assignment of instances located in ambiguous zones.

This result suggests the need to adjust or fine-tune the membership functions. By doing so, the performance of the fuzzy system could be improved, ensuring that all patients are appropriately classified and thereby maximizing the model's accuracy.

5 Conclusions and Future Work

In conclusion, the Sugeno Type-1 Fuzzy System developed using trapezoidal and Gaussian membership functions has proven effective in classifying the risk of heart rate level based on patients' age and heart rate. The results indicate that the model with trapezoidal functions correctly classified 100% of the patients, while the model with Gaussian functions achieved a 93.33% accuracy rate, with only two incorrect classifications. Additionally, in the latter model, 6.67% of patients were not classified into any category, highlighting areas for future improvements.

To strengthen these results, future work is proposed to fine-tune the Gaussian membership functions in heart rate classification. Furthermore, investigating the implementation of Type-2 fuzzy systems could enhance the management of uncertainties and offer greater classification accuracy. Moreover, combining neuro-fuzzy approaches and integrating new clinical data will enhance the system's robustness and its application to a broader population. Finally, metaheuristics could also be used [33–36].

Acknowledgements The authors would like to express thank to the Consejo Nacional de Ciencia y Tecnologia and Tecnologico Nacional de Mexico/Tijuana Institute of Technology for the facilities and resources granted for the development of this research.

References

1. Campero-Jurado, I., Robles-Camarillo, D., Ruiz-Vanoye, J. A., Xicoténcatl-Pérez, J. M., Díaz-Parra, O., Salgado-Ramírez, J. C., Marroquín-Gutiérrez, F., & Ramos-Fernández, J. C. (2024). Fuzzy logic prediction of hypertensive disorders in pregnancy using the Takagi-Sugeno and c-means algorithms. *Mathematics, 12*(15), 1–18.
2. Arif, I. R. F., Firdausi, A., & Hakim, G. P. N. (2021). Nebulizer operational time control based on drug volume and droplet size using fuzzy Sugeno method. *Journal of Robotics and Control (JRC), 2*(2), 94–97.
3. Gaino, R., Covacic, M. R., Cardim, R., Sanches, M. A. A., De Carvalho, A. A., Biazeto, A. R., & Teixeira, M. C. M. (2020). Discrete Takagi-sugeno fuzzy models applied to control the knee joint movement of paraplegic patients. *IEEE Access, 8*, 32714–32726.
4. Moreno-Saenz, J., Chen, Y. J., Tanaka, K., Aragon, J. L., & Quiroz-Juarez, M. A. (2023). A Takagi-Sugeno fuzzy-model-based tracking framework to regulate heart rhythm dynamics. *IEEE Access, 11*, 47223–47235.
5. Dubey, P., Kumar, S., Behera, S. K., & Mishra, S. K. (2023). A Takagi-Sugeno fuzzy controller for minimizing cancer cells with application to androgen deprivation therapy. *Healthcare Analytics, 4*, 1–12.
6. Mariadoss, S., & Augustin, F. (2023). Enhanced Sugeno fuzzy inference system with fuzzy AHP and coefficient of variation to diagnose cardiovascular disease during pregnancy. *Journal of King Saud University—Computer and Information Sciences, 35*(8), 1–25.
7. Meng, X., & Zhang, J. (2020). Anxiety recognition of college students using a Takagi-Sugeno-kang fuzzy system modeling method and deep features. *IEEE Access, 8*, 159897–159905.
8. Ontiveros-Robles, E., Melin, P., & Castillo, O. (2019). Relevance of polynomial order in Takagi-Sugeno fuzzy inference systems applied in diagnosis problems. In *2019 IEEE International Conference on Fuzzy Systems (FUZZ-IEEE)* (pp. 1–6). New Orleans.
9. Ontiveros-Robles, E., Melin, P., & Castillo, O. (2020). Study of the relevance of polynomial order in Takagi-Sugeno Fuzzy inference systems applied in diagnosis problems. In O. Castillo & P. Melin (Eds.), *Hybrid intelligent systems in control, pattern recognition and medicine* (pp. 19–33). Springer International Publishing.
10. Melin, P., & Sánchez, D. (2021). Optimal design of type-2 fuzzy systems for diabetes classification based on genetic algorithms. *International Journal of Hybrid Intelligent Systems, 17*, 15–32.

11. Zhang, L., Zong, G., Zhao, X., Zhao, N., & Sharaf, S. (2023). Reachable set control for discrete-time Takagi-Sugeno Fuzzy singular Markov jump system. *IEEE Transactions on Fuzzy Systems, 31*(9), 3173–3184.
12. Zhang, J., Wang, X., & Shao, X. (2020). Design and real-time implementation of Takagi-Sugeno fuzzy controller for magnetic levitation ball system. *IEEE Access, 8*, 38221–38228.
13. Ma, Y., Wang, Z., Meslem, N., Raissi, T., & Shen, Y. (2023). An improved Zonotopic Approach applied to fault detection for Takagi-Sugeno fuzzy systems. *IEEE Transactions on Fuzzy Systems, 31*(11), 3762–3774.
14. Oh, G. C., & Cho, H. J. (2020). *Blood pressure and heart failure*. BioMed Central Ltd.
15. Schutte, A. E., Kollias, A., & Stergiou, G. S. (2022). Blood pressure and its variability: Classic and novel measurement techniques. *Nature Research*.
16. Muntner, P., Shimbo, D., Carey, R. M., Charleston, J. B., Gaillard, T., Misra, S., Myers, M. G., Ogedegbe, G., Schwartz, J. E., Townsend, R. R., et al. (2019). Measurement of blood pressure in humans: A scientific statement from the American heart association. *Hypertension, 73*(5), e35–e66.
17. Stergiou, G. S., Palatini, P., Parati, G., O'Brien, E., Januszewicz, A., Lurbe, E., Persu, A., Mancia, G., & Kreutz, R. (2021). 2021 European Society of Hypertension practice guidelines for office and out-of-office blood pressure measurement. *Journal of Hypertension, 39*(7), 1293–1302.
18. Bakris, G. L., & Sorrentino, M. J. (2018). *Hypertension, a companion to Braunwald's heart disease* (3rd ed.). Philadelphia: Elsevier.
19. Mancia, G., Grassi, G., Tsioufis, K.P., Dominiczak, A.F., & Agabiti Rosei, E. (2019). *Manual of hypertension of the European society of hypertension*. CRC Press.
20. Pontremoli, R., Leoncini, G., & Viazzi, F. (2019). Hypertension and the kidney. In *Manual of hypertension of the European society of hypertension* (pp. 19–24). CRC Press.
21. Cifkova, R. (2019). Arterial Hypertension and Cardiovascular Risk. In E. A. Andreadis & C. Geladari (Eds.), *Management of hypertension: Current practice and the application of landmark trials* (Vol. Papademetriou, pp. 57–74). Springer International Publishing.
22. Papademetriou, V., Andreadis, E. A., & Geladari, C. (2019). *Management of hypertension*. Springer International Publishing AG.
23. American Heart Association. (2024). *All about heart rate*. American Heart Association. Accessed 20 Apr 2024 [Online]. Available: https://www.heart.org/en/health-topics/high-blood-pressure/the-facts-about-high-blood-pressure/all-about-heart-rate-pulse
24. American Heart Association. (2024). *Target heart rates chart*. American Heart Association. Accessed 20 Apr 2024 [Online]. Available: https://www.heart.org/en/healthy-living/fitness/fitness-basics/target-heart-rates
25. Tiwari, R., Kumar, R., Malik, S., Raj, T., & Kumar, P. (2021). Analysis of heart rate variability and implication of different factors on heart rate variability. *Current Cardiology Reviews, 17*(5), 1–10.
26. Olshansky, B., Ricci, F., & Fedorowski, A. (2023). Importance of resting heart rate. *Trends in Cardiovascular Medicine, 33*(8), 502–515.
27. Palatini, P. (2019). Heart rate as a cardiovascular risk factor in hypertension. In *Manual of hypertension of the European society of hypertension* (pp. 121–126). CRC Press.
28. Avram, R., Tison, G. H., Aschbacher, K., Kuhar, P., Vittinghoff, E., Butzner, M., Runge, R., Wu, N., Pletcher, M. J., Marcus, G. M., et al. (2019). Real-world heart rate norms in the Health eHeart study. *npj Digital Medicine, 2*(1), 1–10.
29. Melin, P., Miramontes, I., & Prado-Arechiga, G. (2018). A hybrid model based on modular neural networks and fuzzy systems for classification of blood pressure and hypertension risk diagnosis. *Expert Systems with Applications, 107*, 146–164.
30. Carvajal, O., Melin, P., Miramontes, I., & Prado-Arechiga, G. (2021). Optimal design of a general type-2 fuzzy classifier for the pulse level and its hardware implementation. *Engineering Applications of Artificial Intelligence*, *97*, 1–7.
31. Melin, P., Prado-Arechiga, G., Miramontes, I., & Guzmán, J. C. (2019). Hypertension diagnosis with a soft computing model using a graphical user interface. *Journal of Hypertension, 37*, Article e233.

32. Miramontes, I., Guzman, J. C., Melin, P., & Prado-Arechiga, G. (2018). Optimal design of interval type-2 fuzzy heart rate level classification systems using the bird swarm algorithm. *Algorithms, 11*(12), 1–35.
33. González, B., Valdez, F., Melin, P., & Prado-Arechiga, G. (2015). Fuzzy logic in the gravitational search algorithm for the optimization of modular neural networks in pattern recognition. *Expert Systems with Applications, 42*(14), 5839–5847.
34. Santiago, A., Dorronsoro, B., Nebro, A. J., Durillo, J. J., Castillo, O., & Fraire, H. J. (2019). A novel multi-objective evolutionary algorithm with fuzzy logic based adaptive selection of operators: FAME. *Information Sciences, 471*, 233–251.
35. Martínez-Soto, R., Castillo, O., & Aguilar, L. T. (2014). Type-1 and Type-2 fuzzy logic controller design using a Hybrid PSO–GA optimization method. *Information Sciences, 285*, 35–49.
36. Castillo, O., Melin, P., Ontiveros, E., Peraza, C., Ochoa, P., Valdez, F., & Soria, J. (2019). A high-speed interval type 2 fuzzy system approach for dynamic parameter adaptation in metaheuristics. *Engineering Applications of Artificial Intelligence, 85*, 666–680. https://doi.org/10.1016/j.engappai.2019.07.020

Interval Type-3 Fuzzy Harmony Search Algorithm for Optimizing Benchmark Functions

Cinthia Peraza, Patricia Ochoa, and Oscar Castillo

Abstract Fuzzy logic theory has been used in different areas of study over the years, starting in engineering and science in the first place, being a crucial part of industrial control and automation. These kinds of theories allow us to transmit human perceptions into computational or electronic systems in real processes. This paper presents the terms and concepts of interval type-3 fuzzy logic (IT3FS) and its application in the dynamic parameter tuning of the fuzzy harmony search algorithm (IT3-FHS) applied to the optimization of mathematical reference functions. The results obtained are shown and a comparison with other existing methods is made in order to validate the efficiency of the proposed method.

Keywords Interval type-3 fuzzy logic · Fuzzy harmony search algorithm · Dynamic parameter adaptation

1 Introduction

The concept of fuzzy refers to something confusing, imprecise or lacking clarity, however, in this area of science it is important to consider it as an adjective, that is, we will refer to those systems as fuzzy because of their control methods and algorithms that are developed to solve various problems, not that these systems are not precise. Fuzzy theory and its concepts were first introduced in engineering by Professor Lotfi Asker Zadeh in 1965 [1]. In his first articles where he addressed fuzzy theory, he explains how linguistic expressions and decisions can be worked

C. Peraza (✉) · P. Ochoa · O. Castillo
Division of Graduate Studies, Tijuana Institute of Technology, Tijuana, México
e-mail: cinthia.peraza@tectijuana.edu.mx

P. Ochoa
e-mail: martha.ochoa18@tectijuana.edu.mx

O. Castillo
e-mail: ocastillo@tectijuana.mx

O. Castillo and P. Melin (eds.), *Intelligent System Design Based on Soft Computing Models*, Studies in Computational Intelligence 1223,
https://doi.org/10.1007/978-3-031-97309-3_2

on mathematically by appreciation, just as we humans do. This theory aims to give a degree or level of belonging within one set or another, instead of the classic sets where an element simply belongs or does not belong. In the real world, systems are not usually linear, nor do we have the mathematical equations for all our systems, electrical, hydraulic, control, among others, to name a few examples. In most cases, there are uncertainties, unknown models, and ambiguity, which is why these systems became so popular in solving problems of this type.

Fuzzy systems have been of great relevance today, applied in different areas and articles of daily or industrial use. In general, we see them in signal processing, communications, expert systems for businesses of different kinds, medicine, and many more.

The present work focuses on the use of IT3FS applied to the Original Harmony Search Algorithm (HS), specifically for the dynamic modification of the HMR parameter. This algorithm [2–5], inspired by the process of musical improvisation, is a metaheuristic optimization technique widely used in complex optimization problems, where the goal is to find an optimal solution by combining "harmony memory" and the generation of new solutions based on controlled variations. In previous studies, the HMR parameter has been tuned using type-1 and type-2 fuzzy logic. These approaches have proven effective in improving the algorithm's adaptability to different types of problems by allowing for more flexible tuning of HS parameters.

However, the use of IT3FS is a more recent approach in the scientific literature [6–11]. This variant of fuzzy logic expands upon the concepts of type-1 and type-2 logics by incorporating a more complex representation of uncertainty, allowing for the modeling of degrees of ambiguity and variability with even greater precision. IT3FS can capture a broader range of uncertainties and provide more robust control in complex systems where uncertainties are not only imprecise but also ambiguous and variable over time.

The main objective of this work is to carry out a first phase of experimentation using a set of mathematical functions, employing IT3FS to validate its behavior and performance in the dynamic modification of the HMR parameter within the HS. This research aims to offer a new perspective on adaptive optimization in metaheuristic algorithms, thus advancing the use of more sophisticated fuzzy logic approaches in this field.

The paper is organized as follows: Sect. 2 provides an overview of the HS, highlighting its key mechanisms and optimization principles. In Sect. 3, we introduce IT3FS, detailing their structure and advantages in handling uncertainty in complex systems. Section 4 presents the integration of IT3FS into the HS, resulting in the IT3FHS, which aims to improve solution accuracy in uncertain environments. Section 5 covers the Benchmark Function Sets used to evaluate the performance of the proposed algorithm, followed by a thorough analysis of the Experimental Results. Finally, Sect. 6 concludes the paper by summarizing the main findings and discussing potential directions for future research.

2 Original Harmony Search Algorithm

Among the existing metaheuristics in the state of the art, this article presents a modification to the HS. HS was first presented by the author Zong Woo Geem in 2001 [12] and since then it has become a widely used optimization algorithm to solve different problems such as: engineering problems [4, 13], parameter optimization [2–4, 14], routing problems [15], medical diagnosis [16], among others. HS is inspired by jazz improvisation imitating the behavior that a musician can have to play a musical piece, the first operation within the improvisation is the use of the harmony memory (HMR) where a musician can play a piece that he has in his memory, the second operation is the pitch adjustment, the musician begins to play a musical piece and adjusts it as he considers until achieving an acceptable musical piece, finally the third parameter is the randomization that would be to play a musical piece randomly. These three operations are represented mathematically in Eqs. (1–3).

$$HMR \in [0,1] \tag{1}$$

$$X_{New} = X_{old} + bp(2\,rand - 1) \tag{2}$$

$$\begin{aligned} PA\ rate &= P_{Lowerlimit} + P_{Range} \times rand \\ where &= P_{Range} = P_{Upperlimit} - P_{Lowerlimit} \end{aligned} \tag{3}$$

3 Interval Type-3 Fuzzy Logic

IT3FS is an extension of conventional fuzzy logic that aims to address the complexity and uncertainty present in increasingly sophisticated systems [17, 18].

$A^{(3)}$ denoted the mathematical modeling for a IT3FS, $A^{(3)}$ is represented by a membership function (MF). The MF of $\mu_{A^{(3)}}$ is denoted by $\mu_{A^{(3)}}(x, u, v)$ and it is called a type-3 membership function (IT3 MF) of the T3FS. In other words, represented in Eqs. (4–7).

$$\mu_{A^{(3)}}: X \times [0, 1] \times [0, 1] \rightarrow [0, 1] \tag{4}$$

$$A^{(3)} = \{x \in X, u \in U \subseteq [0, 1], v \in V \subseteq [0, 1]\} \tag{5}$$

Equations (6–7) show the postulation of a IT3FS, $A^{(3)}$

$$A^{(3)} = \int_{x \in X} \int_{u \in [0,1]} \int_{v \in [0,1]} \mu_{A^{(3)}}(x, u, v)/(x, u, v) \tag{6}$$

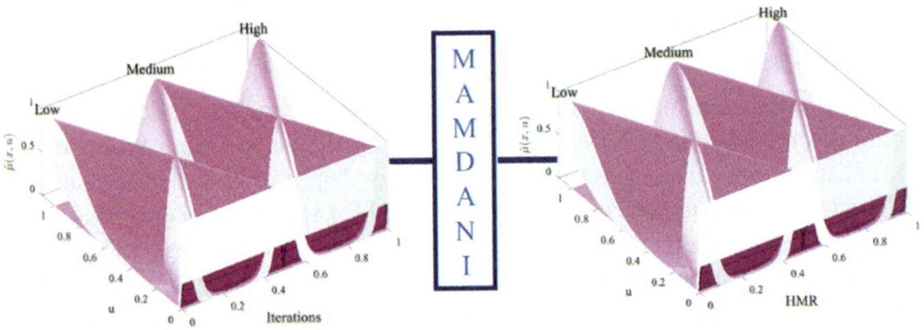

Fig. 1 Interval type-3 fuzzy system

$$A^{(3)} = \int_{x \in X} \left[\int_{u \in [0,1]} \left[\int_{v \in [0,1]} \mu_{A^{(3)}}(x, u, v)/v \right] /u \right] /x \tag{7}$$

4 Interval Type-3 Fuzzy Harmony Search Algorithm

The theory of IT3FS is applied to improve the improvisation of the HS, with this it is intended to dynamically adjust the HMR parameter that represents the exploitation of the search space to achieve the global optimum, therefore the proposal will be known as Interval type-3 fuzzy logic harmony search algorithm (IT3-FHS).

Figure 1 represents the IT3-FHS created for parameter adaptation.

Figure 2 shows the diagram of a process the proposed IT3-FHS.

The approach to dynamically tune the HMR parameter of the IT3-FHS, with the aim of evaluating how this parameter influences the search for solutions of the algorithm. The proposal uses 3 interval-3 triangular membership functions (IT3 MF). Figure 3 shows the visualization of the proposed IT3-FHS (a) which in this case increase as the iterations advance to achieve a better exploitation of the search space, and (b) its corresponding rules.

5 Benchmark Sets of Functions and Experimental Results

The primary goal is to performance review of IT3-FHS using these six benchmark functions, inclunding F1 (Sphere), F2 (Griewank), F3 (Schwefel). F4 (Rastrigin), F5 (Ackley), F6 (Rosenbrock). All these functions were evaluating the results of 10, 30, and 50 dimensions. Figures 4, 5, 6, 7, 8 and 9 presents the graphical representation corresponding to each of these benchmark cases.

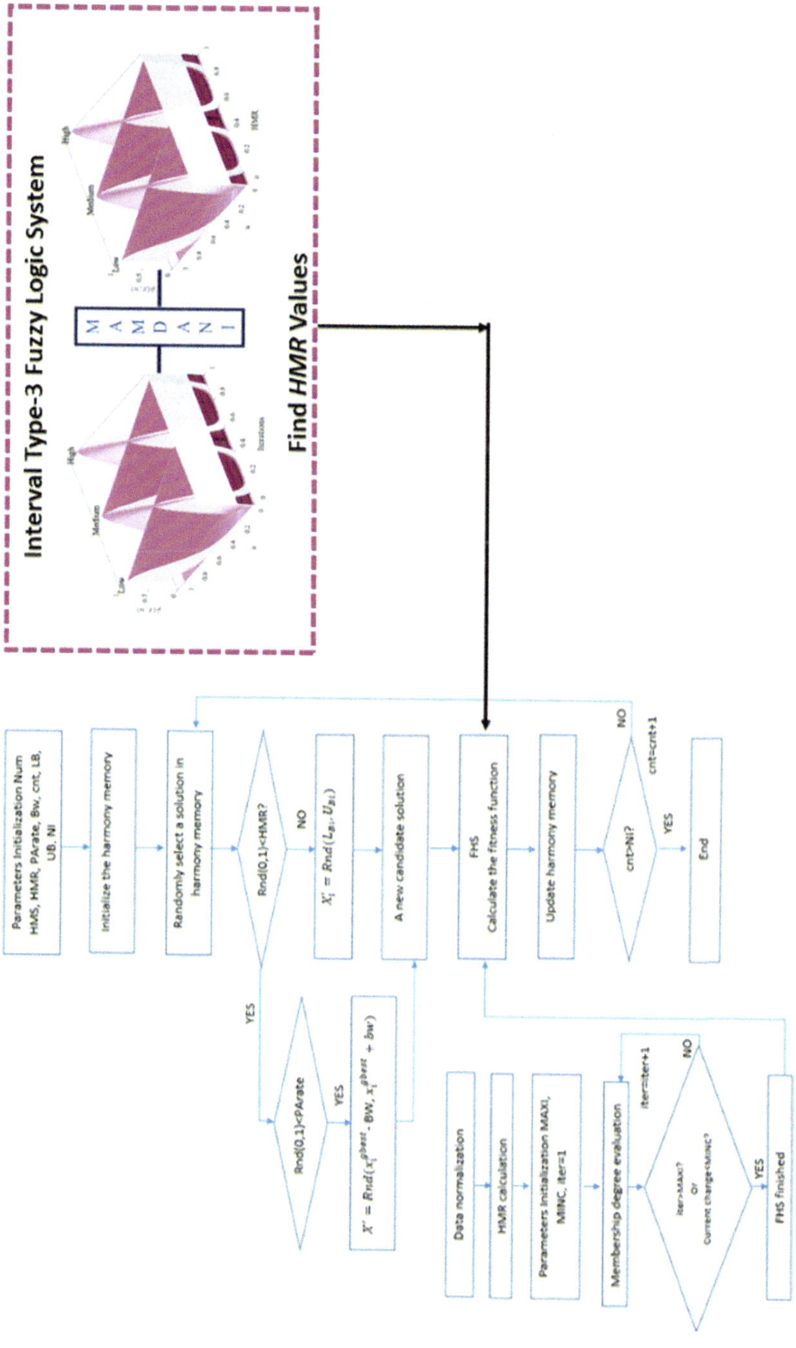

Fig. 2 Diagram of a process in the IT3-FHS

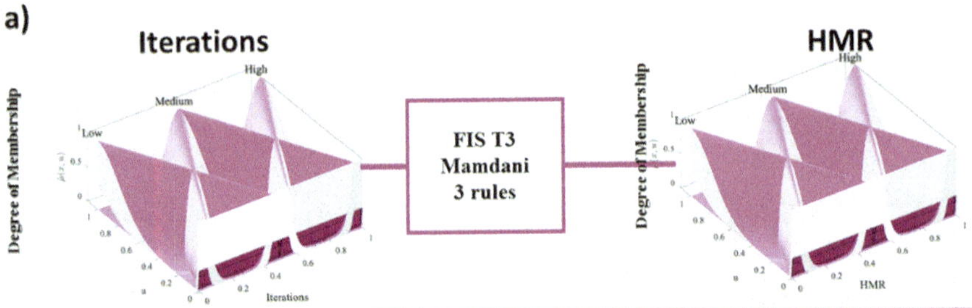

b)

1.- If (Iterations is Low) Then (HMR is Low)
2.-If (Iterations is Medium) Then (HMR is Medium)
3.- If (Iterations is High) Then (HMR is High)

Fig. 3 IT3-FHS adjusting HMR in a increasing manner

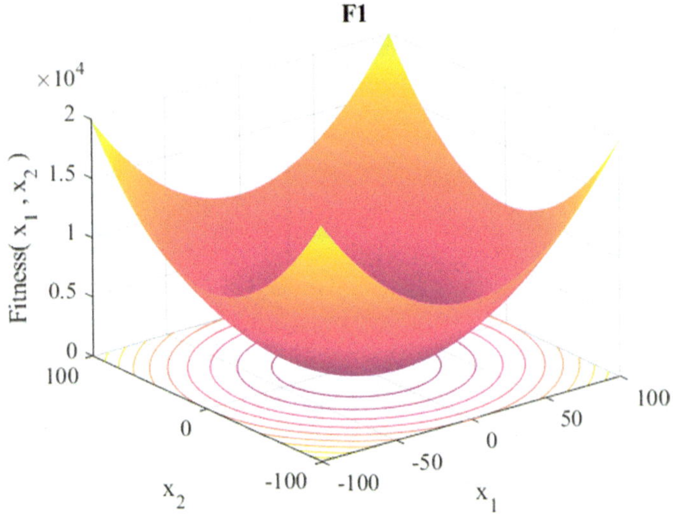

Fig. 4 F1 function

Table 1 shows a comparison with other existing methods, when evaluating 4 functions using 10 dimensions, the T1-FHS method used Type-1 Fuzzy Logic. The average is lower with the T1-FHS method.

Table 2 compares the T2-FHS method with other existing methods. When evaluating 4 functions using 30 dimensions, the T2-FHS method used Interval Type-2 Fuzzy Logic. The average is lower with the proposed method. Therefore, the higher the complexity of the problem, the proposed IT3-FHS reduces the objective.

Table 3 presents the results obtained by increasing the number of dimensions in each mathematical function. The best result is shown in bold, in this case using 50

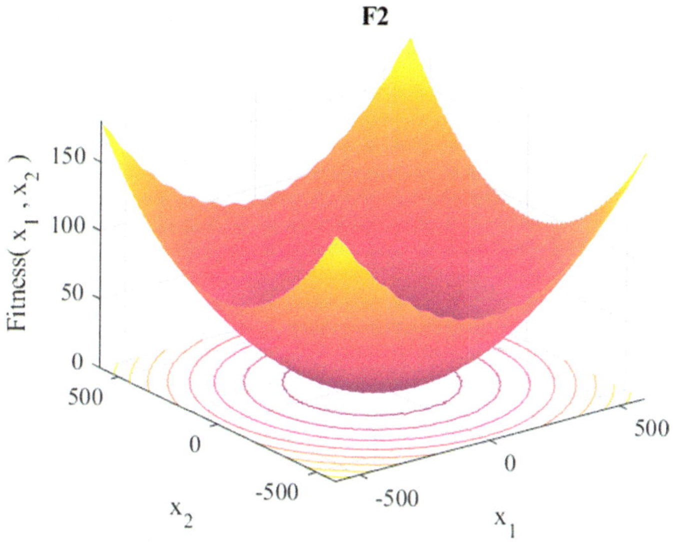

Fig. 5 F2 function

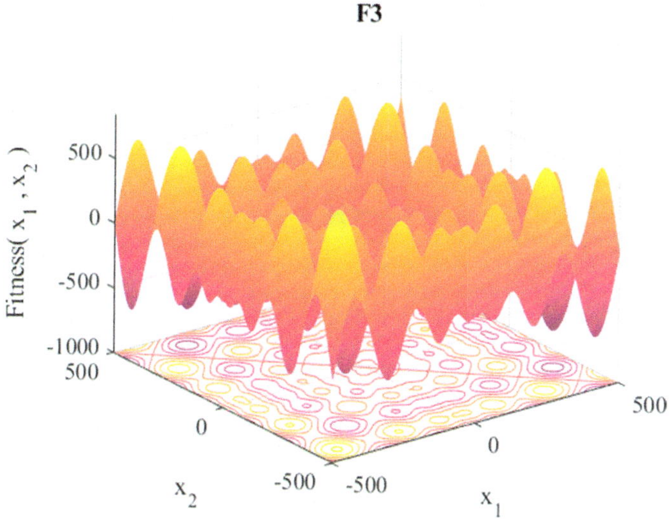

Fig. 6 F3 function

dimensions. We can see that the greater the complexity, the better the error and that the method is efficient when working with greater uncertainty.

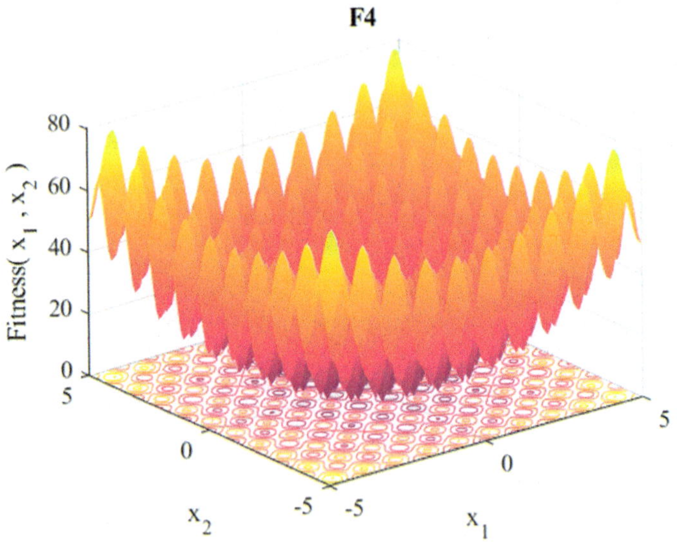

Fig. 7 F4 function

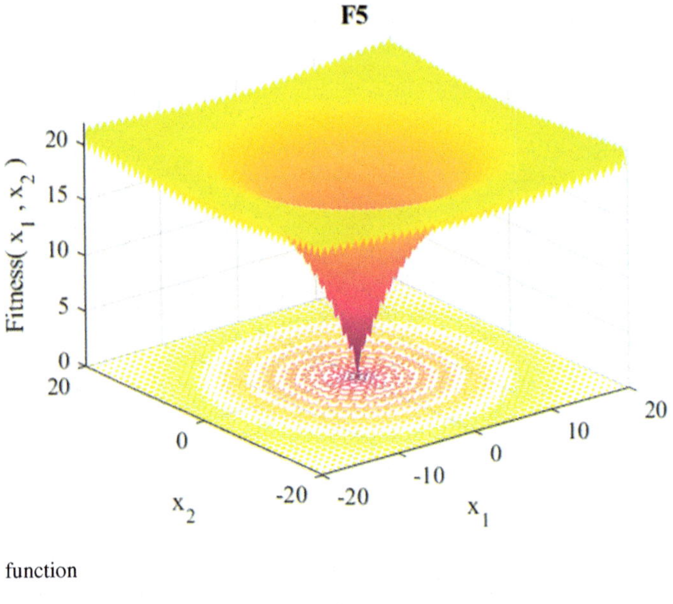

Fig. 8 F5 function

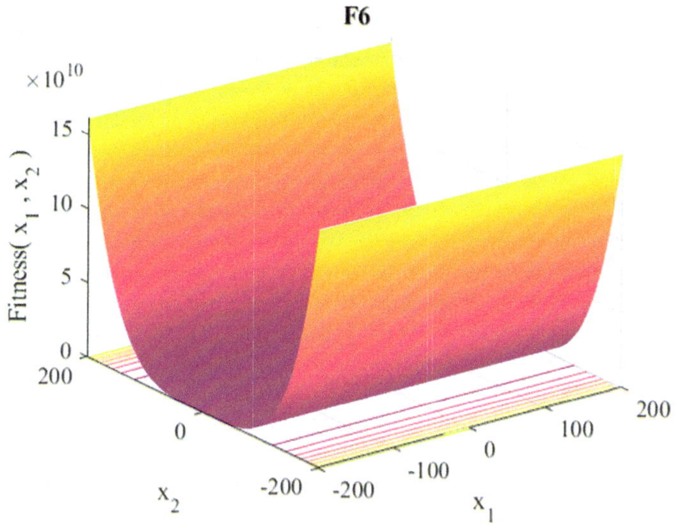

Fig. 9 F6 function

Table 1 Comparison with 10 dimensions

	T1-FHS [19]	IT3-FHS
F1	**6.33E−07**	6.33E−05
F2	–	
F3	–	
F4	**5.28E−05**	4.50E−04
F5	**7.87E−05**	7.77E−05
F6	**1.36E−08**	1.74E−08

Table 2 Comparison with 30 dimensions

	T2-FHS [20]	IT3-FHS
F1	1.44E−10	**1.13E−11**
F2	–	–
F3	–	–
F4	9.55E−06	**5.52E−07**
F5	9.28E−09	**4.38E−10**
F6	2.52E−12	**2.15E−12**

Table 3 Results of the simulation with different dimensions applying the proposed method

		Dimensions		
	Performance index	10	30	50
F1	Mean	6.33E−05	1.13E−11	**3.25E−12**
	SD	2.40E−05	1.13E−11	1.55E−11
	Best result	1.36E−08	1.44E−13	4.22E−15
	Worsr result	9.69E−05	8.46E−11	8.46E−11
F2	Mean	6.72E−01	1.70E−07	**1.29E−08**
	SD	5.87E−01	6.58E−08	1.06E−08
	Best result	2.38E−03	5.67E−08	1.05E−10
	Worsr result	1.80E+00	2.80E−07	5.67E−08
F3	Mean	9.17E−01	3.64E−06	**4.11E−08**
	SD	1.10E+00	2.72E−06	2.70E−08
	Best result	1.69E−01	5.67E−08	5.67E−10
	Worsr result	4.82E+00	8.29E−06	8.62E−08
F4	Mean	4.50E−04	5.52E−07	**2.25E−08**
	SD	1.55E−03	2.81E−07	6.68E−09
	Best result	5.75E−05	5.67E−08	1.15E−08
	Worsr result	8.35E−03	1.01E−06	3.35E−08
F5	Mean	7.77E−05	4.38E−10	4.41E−12
	SD	3.89E−05	2.65E−10	**9.32E−13**
	Best result	1.04E−06	4.66E−12	9.84E−13
	Worsr result	1.29E−04	8.71E−10	4.66E−12
F6	Mean	1.74E−08	2.15E−12	**1.66E−13**
	SD	9.77E−09	7.47E−12	1.69E−13
	Best result	5.42E−09	3.19E−18	3.26E−16
	Worsr result	5.67E−08	3.03E−11	5.38E−13

6 Conclusions

This work introduces an enhancement to the harmony search algorithm by incorporating interval Type-3 fuzzy logic to adjust the HMR parameter dynamically. The architecture of the proposed method was tested on six benchmark functions with varying problem dimensions to validate its performance. In the result analysis, it is observed that for problems with lower complexity specifically those with 10 dimensions the proposed method demonstrates stable results; however, compared to other methods in the literature, there is no significant difference in performance. Conversely, as the problem complexity increases, the method's effectiveness becomes more pronounced. For instance, with 30 and 50 dimensions, the proposed method shows improved efficiency by better managing the uncertainty of the objective function. This capacity to handle greater complexity underscores one of the primary advantages of employing interval Type-3 fuzzy logic: as problem complexity grows, the results become more robust, like in [21–23].

References

1. Zadeh, L. A. (1965). Fuzzy sets. *Information and Control, 8*, 338–353.
2. Karmakar, R., Chatterjee, S., Datta, D., & Chakraborty, D. (2024). Application of harmony search algorithm in optimizing autoregressive integrated moving average: A study on a data set of Coronavirus Disease 2019. *Systems and Soft Computing, 6*, Article 200067.
3. Atta, S. (2024). An improved harmony search algorithm using opposition-based learning and local search for solving the maximal covering location problem. *Engineering Optimization, 56*(8), 1298–1317.
4. Sun, X., He, H., & Ma, L. (2024). Harmony search meta-heuristic algorithm based on the optimal sizing of wind-battery hybrid micro-grid power system with different battery technologies. *Journal of Energy Storage, 75*, Article 109582.
5. Wang, B., Zhang, P., Wang, X., & Pan, Q. (2024). Three-way decision based island harmony search algorithm for robust flow-shop scheduling with uncertain processing times depicted by big data. *Applied Soft Computing, 111842.*
6. Castillo, O., Castro, J. R., & Melin, P. (2022). *Interval type-3 fuzzy systems: Theory and design.* Springer International Publishing.
7. Qasem, S. N., Ahmadian, A., Mohammadzadeh, A., Rathinasamy, S., & Pahlevanzadeh, B. (2021). A type-3 logic fuzzy system: Optimized by a correntropy based Kalman filter with adaptive fuzzy kernel size. *Information sciences, 572*, 424–443.
8. Hamdy, M., Ibrahim, A., Abozalam, B., & Helmy, S. (2024). Design and implementation of Type-3 intuitionistic fuzzy logic MPPT controller for PV solar system: Comparative study. *ISA Transactions.*
9. Tian, M. W., Yan, S. R., Mohammadzadeh, A., Tavoosi, J., Mobayen, S., Safdar, R., & Zhilenkov, A. (2021). Stability of interval type-3 fuzzy controllers for autonomous vehicles. *Mathematics, 9*(21), 2742.
10. Taghieh, A., Mohammadzadeh, A., Zhang, C., Kausar, N., & Castillo, O. (2022). A type-3 fuzzy control for current sharing and voltage balancing in microgrids. *Applied Soft Computing, 129*, Article 109636.
11. Cao, Y., Raise, A., Mohammadzadeh, A., Rathinasamy, S., Band, S. S., & Mosavi, A. (2021). Deep learned recurrent type-3 fuzzy system: Application for renewable energy modeling/prediction. *Energy Reports, 7*, 8115–8127.
12. Geem, Z. W., Kim, J. H., & Loganathan, G. V. (2001). A new heuristic optimization algorithm: Harmony search. *Simulation, 76*, 60–68.
13. Gomez-Gil, F. J., Martínez-Martínez, V., Ruiz-Gonzalez, R., Martínez-Martínez, L., & Gomez-Gil, J. (2024). Vibration-based monitoring of agro-industrial machinery using a k-Nearest Neighbors (kNN) classifier with a Harmony Search (HS) frequency selector algorithm. *Computers and Electronics in Agriculture, 217*, Article 108556.
14. Ouyang, H., Liang, S., Li, S., Zhou, Z., & Zhan, Z. H. (2024). A dual population collaborative harmony search algorithm with adaptive population size for the system reliability-redundancy allocation problems. *Journal of Computational Design and Engineering, 11*(2), 285–313.
15. Minanda, V., Liang, Y. C., Chen, A. H., & Gunawan, A. (2024). Application of an improved harmony search algorithm on electric vehicle routing problems. *Energies, 17*(15), 3716.
16. Zhang, Z., Lu, Y., Ye, M., Huang, W., Jin, L., Zhang, G., & Zhu, W. (2024). A novel evolutionary ensemble prediction model using harmony search and stacking for diabetes diagnosis. *Journal of King Saud University-Computer and Information Sciences, 36*(1), Article 101873.

17. Deveci, M., Simic, V., Karagoz, S., & Antucheviciene, J. (2022). An interval type-2 fuzzy sets based Delphi approach to evaluate site selection indicators of sustainable vehicle shredding facilities. *Applied Soft Computing, 108465.*
18. Mohammadzadeh, A., Castillo, O., Band, S. S., et al. (2021). A novel fractional-order multiple-model type-3 fuzzy control for nonlinear systems with unmodeled dynamics. *International Journal of Fuzzy Systems, 23*, 1633–1651. https://doi.org/10.1007/s40815-021-01058-1
19. Peraza, C., Valdez, F., & Castillo, O. (2015). Fuzzy control of parameters to dynamically adapt the hs algorithm for optimization. In *Fuzzy Information Processing Society (NAFIPS) Held Jointly with 2015 5th World Conference on Soft Computing* (pp. 1–6). IEEE.
20. Peraza, C., Valdez, F., & Castillo, O. (2019). Fuzzy harmony search algorithm using an interval type-2 fuzzy logic applied to benchmark mathematical functions. In *Intuitionistic fuzziness and other intelligent theories and their application* (pp. 13–28).
21. González, B., Valdez, F., Melin, P., & Prado-Arechiga, G. (2015). Fuzzy logic in the gravitational search algorithm for the optimization of modular neural networks in pattern recognition. *Expert Systems with Applications, 42*(14), 5839–5847.
22. Castillo, O., Melin, P., Ontiveros, E., Peraza, C., Ochoa, P., Valdez, F., & Soria, J. (2019). A high-speed interval type 2 fuzzy system approach for dynamic parameter adaptation in metaheuristics. *Engineering Applications of Artificial Intelligence, 85*, 666–680. https://doi.org/10.1016/j.engappai.2019.07.020
23. Amador-Angulo, L., Castillo, O., Melin, P., & Geem, Z. G. (2024). Type-3 fuzzy dynamic adaptation of Bee colony optimization applied to mathematical functions. *Fuzzy Sets and Systems, 489*. https://doi.org/10.1016/j.fss.2024.109014

Analysis Comparative of a Mamdani Versus Takagi-Sugeno Defuzzification Method of a Fuzzy Bee Colony Optimization Algorithm to Stable the Trajectory in an Autonomous Mobile Robot

Leticia Amador-Angulo and Oscar Castillo

Abstract A Takagi-Sugeno system with a Generalized type-2 Fuzzy logic system (SGT2FLS) is utilized to find of parameters in a Bee Colony Optimization (BCO) to control the motion of an autonomous mobile robot (AMR) is presented in this paper. The first objective of BCO algorithm is to find the optimal design of the Membership Functions (MFs) in a Mamdani Generalized Type-2 Fuzzy Logic System (MGT2FLS), the second idea was using a Fuzzy BCO-SGT2FLS to find the optimal values of α and β parameters in BCO and an analysis of the errors obtained by the Mean Square Error (MSE) metric is presented. The experimentations demonstrate that BCO has an excellent execution when a FBCO-SGT2FLS is executed. In addition, a comparative with a FBCO-MGT2FLS is depicted.

Keywords Virtual bee · Takagi-Sugeno inference · Robot autonomous mobile · Stabilization

1 Introduction

In last year, the efficiency and fast that a Takagi-Sugeno has to obtain better results in different type of applications and to solve problems has increased considerably, and several authors have shown their interest in studying the behavior of this inference method in Fuzzy Sets (FSs): Amador-Angulo et al. present a smart Sugeno interval Type-2 FLS in [1], in [16] Mersadek et al. proposal a Takagi-Sugeno fuzzy controller for frequency control, in [17] Moali et al. presents an Type-2 FLC with genetic algorithm, in [18] Nguyen et al. study a Takagi–Sugeno fuzzy applied in autonomous ground vehicles, in [19] Pulido et al. present a comparison of two inference method

L. Amador-Angulo (✉) · O. Castillo
Tijuana Institute of Technology, Tijuana, México
e-mail: gloria.amador@tectijuana.edu.mx

O. Castillo and P. Melin (eds.), *Intelligent System Design Based on Soft Computing Models*, Studies in Computational Intelligence 1223,
https://doi.org/10.1007/978-3-031-97309-3_3

applied to ensemble neural networks, in [20] Tavoosi et al. propose a new general type-2 prediction for PID tuning, in [25] Wang et al. study an IT2 Takagi-Sugeno FS for event-based predictive control, in [27] Zhao et al. analyzed a GT2FS gain scheduling PID controller, and in [28] Zhou et al. study on non-iterative algorithms for center-of-sets type-reduction of IT2 Takagi-Sugeno-Kang FLS. The principal contribution is to find the $\boldsymbol{\alpha}$ and $\boldsymbol{\beta}$ the parameter values in BCO based on a Takagi-Sugeno method of inference of a GT2FLS and the design of the MFs of the T1FLS applied to Fuzzy Logic Controller (FLC) for the trajectory in an AMR.

The hybridization of FSs with bio-inspired algorithm to solve complex problems is a good tool to improve the performance in the original techniques, for example; in [3] Bernal et al. present an Optimization of type-2 fuzzy logic controller design using the GSO and FA algorithms, in [5] Castillo et al. introduce an improve of a differential evolution algorithm with type-2 fuzzy logic, and in [7] Cuevas et al. propose a generalized type-2 fuzzy parameter adaptation in the marine predator algorithm for fuzzy controller applied to problem of robotics.

The statement problem studied in this paper is based in controlling the trajectory in an autonomous mobile robot (AMR) is the interest by several authors; in [10] Galarza-Falfan et al. study a path planning for AMR using intelligent algorithms, in [11] Hassan et al. implement an AMR to planning path and trajectory tracking control, in [12] Liu et al. introduce a review of sensing technologies for indoor AMR, in [23] Top et al. propose a design of an android applications for AMR, and in [24] Vignesh et al. development of rapidly exploring random tree based AMR navigation and velocity predictions using K-nearest neighbors with fuzzy logic analysis. Several techniques have been implemented to find the better trajectory in the AMR, in this paper the main contribution is to find the values of the α and β parameters using a generalized type-2 FLS of Takagi-Sugeno inference method with the goal in controlling the trajectory in an AMR.

The paper is divided as a following; the proposed Sugeno-Takagi Fuzzy Generalized BCO algorithm is outlined in Sect. 2. Section 3 presents in detail the case study. Results and comparative are presented in Sect. 4. Finally, Sect. 5 shows the relevant conclusions found.

2 Bee Colony Optimization Algorithm

2.1 Original BCO Algorithm

Teodorović Dušan in 2006 propose the first idea based on intelligent virtual bee with the Bee Colony Optimization Algorithm (BCO) [22]. The idea is focused in to explore a set of bees with the objective to collect nectar [9]. The BCO is compose by two essential passes: Backward and Forward, in the first pass the idea is initial the exploration and the second pass help in the exploitation [8], a bee during the execution can have three types of roles which as: bee, scout and follower [21]. The dynamic in the BCO is presented in Eqs. (1–4).

$$P_{ij,n} = \frac{\left[\rho_{ij,n}\right]^{\alpha} \cdot \left[\frac{1}{d_{ij}}\right]^{\beta}}{\sum_{j \in A_{i,n}} \left[\rho_{ij,n}\right]^{\alpha} \cdot \left[\frac{1}{d_{ij}}\right]^{\beta}} \tag{1}$$

$$D_i = K \cdot \frac{Pf_i}{Pf_{colony}} \tag{2}$$

$$Pf_i = \frac{1}{L_I}, L_i = Tour\,Length \tag{3}$$

$$Pf_{colony} = \frac{1}{N_{Bee}} \sum_{i=1}^{N_{Bee}} Pf_i \tag{4}$$

Each parameter is defined in previous research [1, 2, 6]. A virtual bee presents a movement when a good solution is found called "waggle dance" [4], and Fig. 1 shown the flowchart of the original BCO.

2.2 Proposal of Fuzzy BCO Algorithm

Lofti Zadeh in 1965 propose the first idea in the T1FLS [26], Mendel et al. in 2006 study an extension of the T1FLS with IT2FLS [15], and finally, an efficiency tools in the evaluation of the uncertainty is proposal with the GT2FLS by Mendel et al. in 2013 [13] to formulate a GT2 FS $\tilde{\tilde{A}}$ is formulated by Eq. (5):

$$\tilde{\tilde{A}} = \left\{\left((x, u), \mu_{\tilde{A}}(x, u)\right) | \forall x \in X, \forall u \in J_x \subseteq [0, 1]\right\} \tag{5}$$

where $J_x \subseteq [0,1]$, x is the partition of the primary MF, and u is the partition of the secondary MF. Figure 2 depicts the representation of GT2MFs, and in Fig. 3, the footprint of uncertainty (FOU) is highlighted, which is better to evaluate the uncertainty in real problems. In this case a GT2FS uses $f_x(u)$, in the vertical axis. Figure 4 show the general structure of a GT2FLS.

The α-plane for a GT2FLS, in this case $\tilde{A}$, is denoted by $\tilde{A}\alpha$, and it is the union of all primary MFs of $\tilde{A}$, which secondary membership degrees are higher or equal to α ($0 \leq \alpha \leq 1$) [14]. The mathematical representation is expressed by Eq. (6). In Fig. 5 the representation of an α-Planes is illustrated [14].

$$\tilde{A}_{\alpha} = \left\{(x, u), \mu_{\tilde{A}}(x, u) \geq \alpha | \forall x \in X, \forall u \in J_X \subseteq [0,1]\right\} \tag{6}$$

The proposed methodology in created in [1, 2, 6] in this case the objective is to use a Takagi–Sugeno inference method of a GT2FLS. The design is composed by two inputs; iterations and diversity, Eqs. (7) and (8) indicate the mathematical representation to find the values in each execution for the proposal method [2].

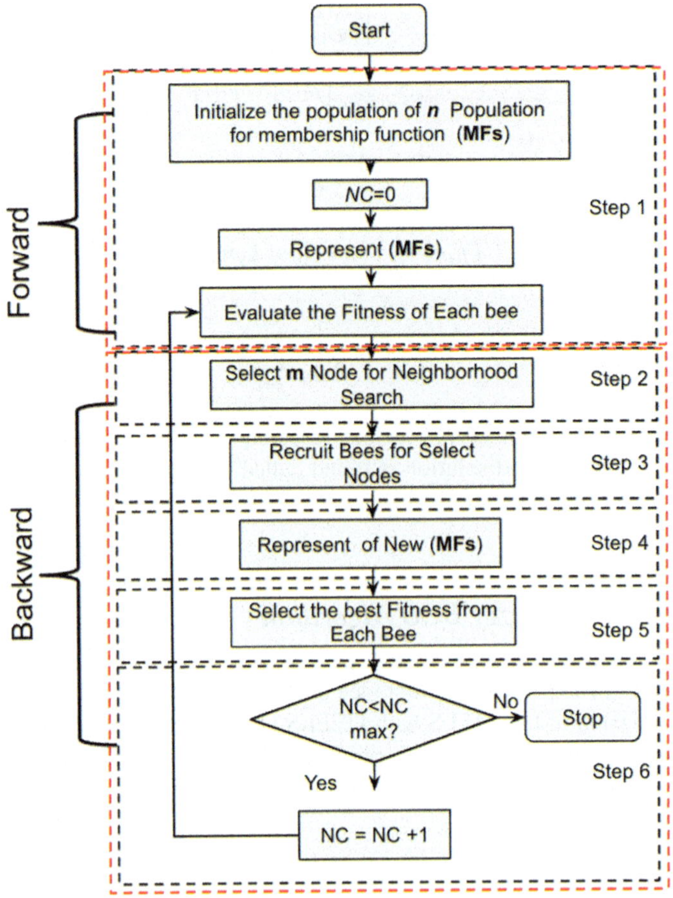

Fig. 1 Steps of BCO

Fig. 2 Visual representation of a triangular GT2 MF

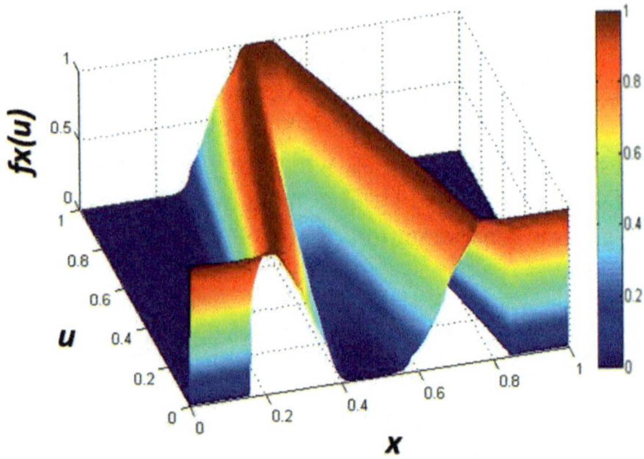

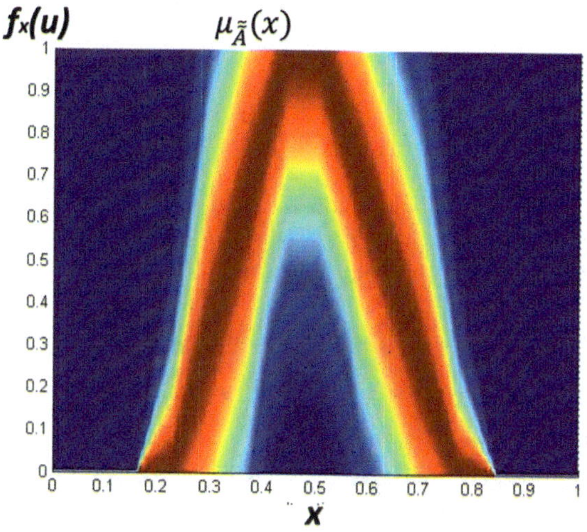

Fig. 3 FOU of the GT2MF

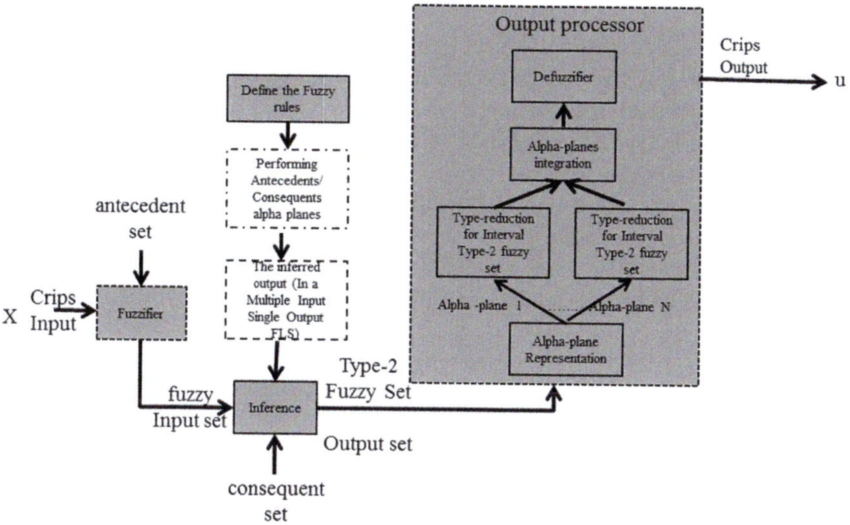

Fig. 4 Structure of a GT2FLS

$$\text{Iteration} = \frac{Current\ Iteration}{Maximum\ of\ Iterations} \tag{7}$$

$$Diversity(S(t)) = \frac{1}{n_s}\sum_{i=1}^{n_x}\sqrt{X_{ij}(t) - \overline{X}_j(t)^2} \tag{8}$$

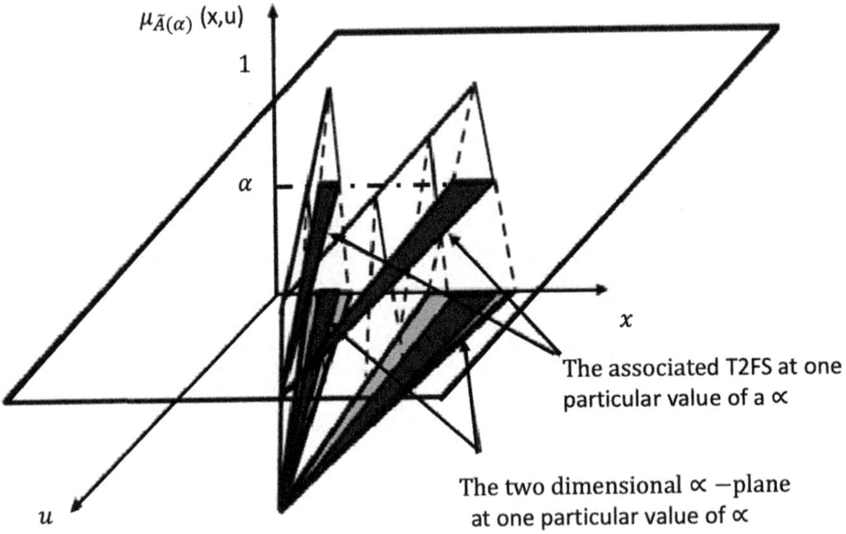

Fig. 5 A visualization of the associated type-2 FS for ∝-plane

Based on Eq. (8), the diversity indicates the distance degree of the bees. Each parameter is expressed in Table 1.

The proposal Takagi–Sugeno FBCO- GT2FLS has two inputs; iteration and diversity with three triangular MFs, and two outputs, where the linguistic variables of the antecedents are granulated into low, mediumlow, medium, mediumhigh and high linguistic values. The consequents of this GT2FLS consist of constants, which are Alpha (α) and Betha (β), which are exhibited in Fig. 6. The fuzzy rules designed is shown in Figs. 7 and 8 depicts the surface of the SGT2FLS.

Table 1 Representation of FBCO-SGT2FLS

Parameter	Representation
t	Current iteration
n_s	Population size
i	A bee
n_x	Total of solutions
j	Next solution
X_{ij}	Solution of j
$\underline{X}_j$	Solution of j in the best bee

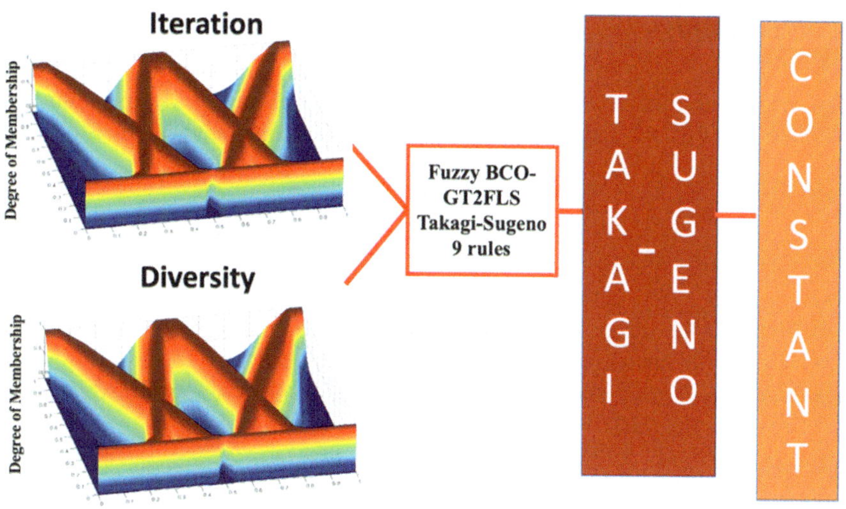

Fig. 6 General structure for the FBCO-SGT2FLS

Fig. 7 Fuzzy rules in the SFBCO-GT2FLS

# Rules	Iteration	Diversity	Beta	Alpha
	Inputs		Outputs	
1	Low	Low	High	Low
2	Low	Medium	MediumHigh	Medium
3	Low	High	MediumHigh	MediumLow
4	Medium	Low	MediumHigh	MediumLow
5	Medium	Medium	Medium	Medium
6	Medium	High	MediumLow	MediumHigh
7	High	Low	Medium	High
8	High	Medium	MediumLow	MediumHigh
9	High	High	Low	High

Fig. 8 Surface of the SGT2FLS for two outputs

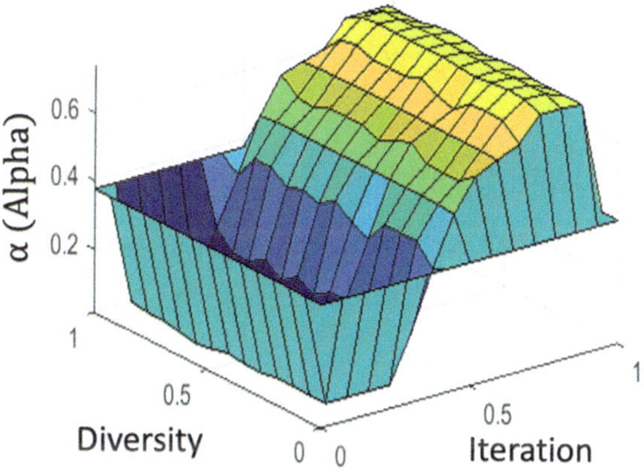

3 Case Study

The model describes a unicycle mobile robot [1, 2, 6], Fig. 9 illustrates this real problem, and Eqs. (9) and (10) expresses the representation mathematical.

$$M(q)\dot{v} + C(q, \dot{q})v + Dv = \tau + P(t) \tag{9}$$

where, $q = (x, y, \theta)^T$ is the coordinates vector, $\upsilon = (v, w)^T$ is the velocities vector, $\tau = (\tau_1, \tau_2)$ is the vector where τ_1 (right) and τ_2 (left) represent the torques, $P \in R^2$ is a uniformly bounded vector of disturbances, $M(q) \in R^{2\times 2}$ is the positive-definite inertia matrix, $C(q, \dot{q})^{\vartheta}$ is the vector of Coriolis and centripetal forces, and $D \in R^{2\times 2}$ is expressed in Eq. (10).

$$\dot{q} = \underbrace{\begin{bmatrix} \cos\theta & 0 \\ \sin\theta & 0 \\ 0 & 1 \end{bmatrix}}_{J_q} \underbrace{\begin{bmatrix} v \\ w \end{bmatrix}}_{v} \tag{10}$$

where, (x, y) are the positions of reference frame, x-axis indicates the v (linear velocity) and w (angular velocity) in the θ is the angle. Equation (11) postulates the relation to a non-slip wheel condition.

$$\dot{y}\, cos\theta - \dot{x}\, sin\theta = 0 \tag{11}$$

The T1FLS is designer with two inputs: the first is LE (linear error), with three MFs with fuzzy sets of N, Z and P. The second is WE (angular error) with three MFs. The T1FLS has two outputs called T1 (Torque 1), and T2 (Torque 2), which are composed of three triangular MFs with the following linguistic values, respectively: N, Z, P. The T1FLS is illustrated in Fig. 10 and the combination of the rules is presented in Fig. 11, and their description can be found in [1] and Fig. 12 shows the model.

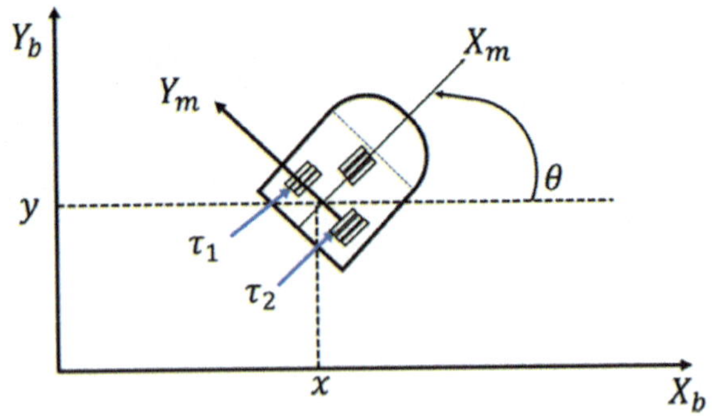

Fig. 9 Robot representation

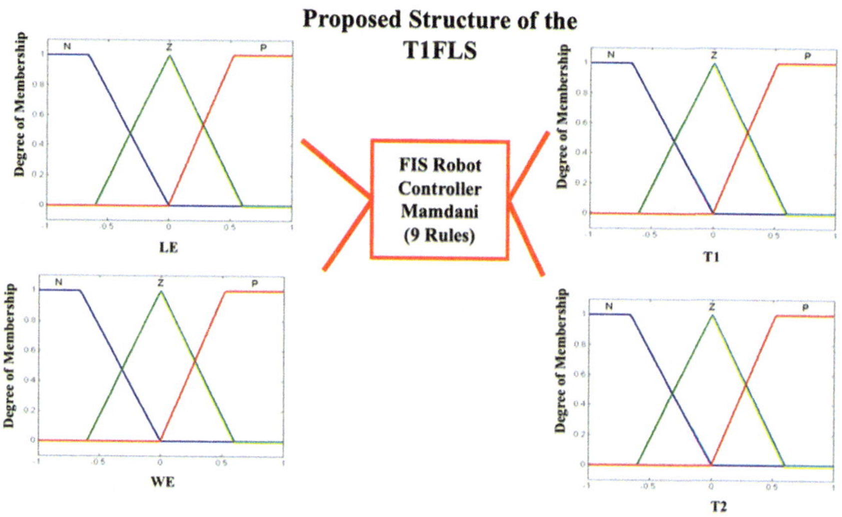

Fig. 10 General Structure of the T1FLS

Fig. 11 Rules of the T1FLS

# Rules	LE	WE	T2	T1
1	N	N	N	N
2	N	Z	N	Z
3	N	P	N	P
4	Z	N	Z	N
5	Z	Z	Z	Z
6	Z	P	Z	P
7	P	N	P	N
8	P	Z	P	Z
9	P	P	P	P
	Inputs		Outpus	

4 Results

Table 2 lists the values for each important parameter in the FBCO-SGT2FLS.

The function selected by analyzed the performance is the MSE denoted in Eq. (12), and Table 3 detailed a comparative with two types of systems.

$$MSE = \frac{1}{n}\sum_{i=1}^{n}\left(\overline{Y}_i - Y_i\right)^2 \quad (12)$$

Results on Table 2 indicates that when the proposed method is implemented the minimal errors is found in comparison with the original BCO and Mamdani inferences system FBCO-GT2FLS [1]. Figure 13 illustrates a visualization of the

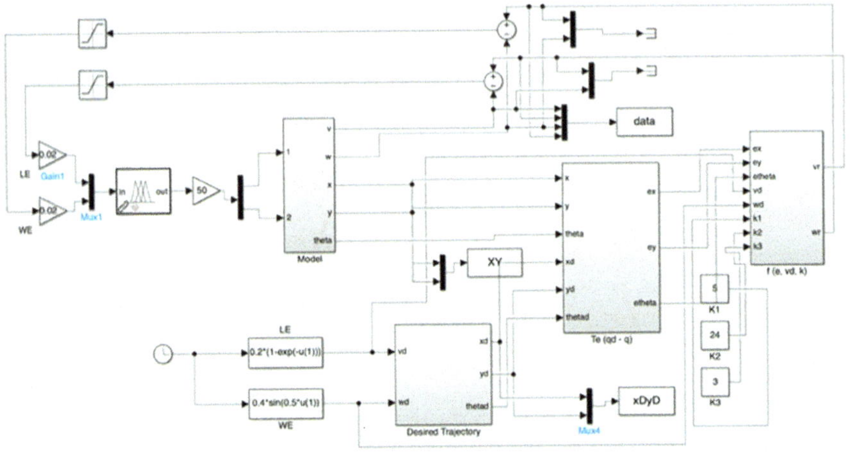

Fig. 12 Model for the problem

Table 2 Parameter setting for FBCO-SGT2FLS

Parameters	Methods	
	Original BCO	FBCO-SGT2FLS
Population	50	50
Follower Bee	25	25
Iterations	30	30
α	0.5	Dynamic
β	2.5	Dynamic

Table 3 Comparative of Mamdani and Takagi-Sugeno FBCO-IT2FLS

Metrics	*Methods*		
	Original BCO	*Mamdani FBCO-GT2FLS* [2]	*Sugeno FBCO-GT2FLS*
MSE(Best)	1.01E-01	3.00E-03	**3.70E-02**
Worst	2.84E-01	6.65E + 01	1.27E + 01
Average	1.96E-01	6.12E + 00	1.88E + 00
σ	4.66E-02	1.30E + 01	2.84E-01
Alpha	0.5	0.466	0.484
Betha	2.5	2.599	2.501

convergence for the best results obtained by proposal method, and the best trajectory is depicted in Fig. 14.

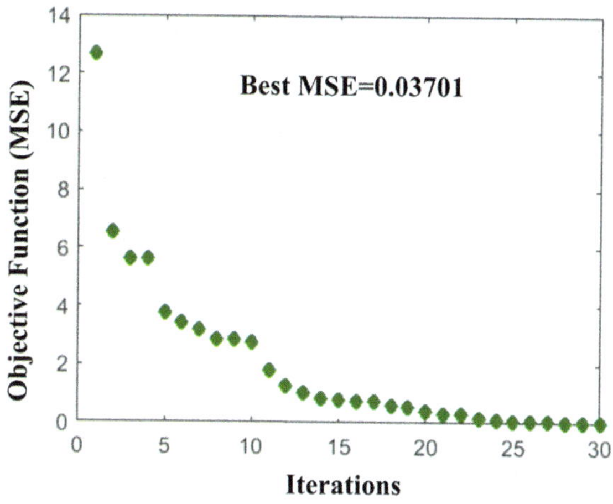

Fig. 13 Convergence of the best results found by FBCO-SGT2FLS

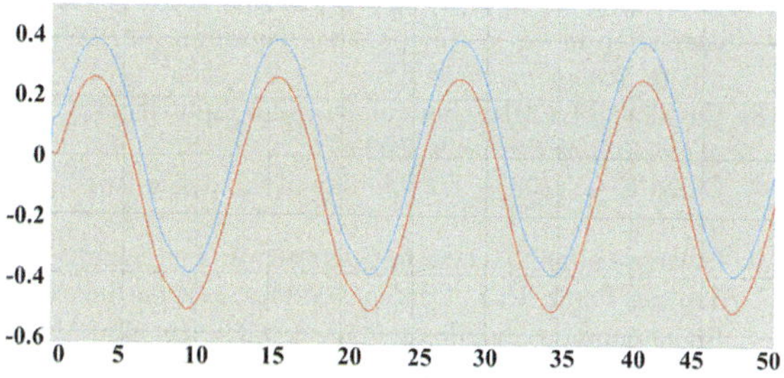

Fig. 14 Representation of the best trajectory found by FBCO-SGT2FLS

Figure 14 depicts the best trajectory found by proposal method with a MSE of 0.0370, blue line indicates the desired trajectory and red line is the trajectory found by FBCO-GT2FLS.

5 Conclusions

The main conclusion in analyzing the stabilization of the AMR using a Takagi-Sugeno fuzzy system, with the goal to find the adequate values in the α and β parameters, when compared to a Mamdani FBCO-GT2FLS is that Sugeno produces the better results (See Table 3). The experiments determine that the proposed algorithm is better in results when compared to the BCO algorithm (Fig. 13). The values to α and β parameters that FBCO-SGT2FLS recommend are 0.484 and 2.501, respectively (see Table 3). The trajectory found by the proposed method is closer to the desired trajectory (Fig. 14).

As future work, FBCO-GT2FLS will be implemented in the optimization of GT2FLC. A Takagi–Sugeno Interval T3FLS using other types of MFs is an interesting future works [29].

References

1. Amador-Angulo, L., & Castillo, O. (2024). A smart sugeno interval type-2 fuzzy bee colony optimization to stable an autonomous mobile robot controller. In *International conference on intelligent and fuzzy systems* (pp. 580–588). Springer Nature Switzerland.
2. Amador-Angulo, L., Mendoza, O., Castro, J. R., Rodríguez-Díaz, A., Melin, P., & Castillo, O. (2016). Fuzzy sets in dynamic adaptation of parameters of a bee colony optimization for controlling the trajectory of an autonomous mobile robot. *Sensors, 16*(9), 1458.
3. Bernal, E., Lagunes, M. L., Castillo, O., Soria, J., & Valdez, F. (2021). Optimization of type-2 fuzzy logic controller design using the GSO and FA algorithms. *International Journal of Fuzzy Systems, 23*(1), 42–57.
4. Biesmeijer, J. C., & Seeley, T. D. (2005). The use of waggle dance information by honey bees throughout their foraging careers. *Behavioral Ecology and Sociobiology, 59*(1), 133–142.
5. Castillo, O., Ochoa, P., & Soria, J. (2020). *Differential evolution algorithm with type-2 fuzzy logic for dynamic parameter adaptation with application to intelligent control*. Springer Nature.
6. Castillo, O., Amador-Angulo, L., Castro, J. R., & Garcia-Valdez, M. (2016). A comparative study of type-1 fuzzy logic systems, interval type-2 fuzzy logic systems and generalized type-2 fuzzy logic systems in control problems. *Information Sciences, 354*, 257–274.
7. Cuevas, F., Castillo, O., & Cortés-Antonio, P. (2022). Generalized type-2 fuzzy parameter adaptation in the marine predator algorithm for fuzzy controller parameterization in mobile robots. *Symmetry, 14*(5), 859.
8. Davidović, T. (2016). Bee colony optimization Part I: The algorithm overview. *Yugoslav Journal of Operations Research, 25*(1).
9. Dyler, F. C. (2002). The biology of the dance language. *Annual Review of Entomology, 47*, 917–949.
10. Galarza-Falfan, J., García-Guerrero, E. E., Aguirre-Castro, O. A., López-Bonilla, O. R., Tamayo-Pérez, U. J., Cárdenas-Valdez, J. R., & Inzunza-Gonzalez, E. (2024). Path planning for autonomous mobile robot using intelligent algorithms. *Technologies, 12*(6), 82.
11. Hassan, I. A., Abed, I. A., & Al-Hussaibi, W. A. (2024). Path planning and trajectory tracking control for two-wheel mobile robot. *Journal of Robotics and Control (JRC), 5*(1), 1–15.
12. Liu, Y., Wang, S., Xie, Y., Xiong, T., & Wu, M. (2024). A Review of sensing technologies for indoor autonomous mobile robots. *Sensors, 24*(4), 1222.
13. Mendel, J. M. (2013). General type-2 fuzzy logic systems made simple: A tutorial. *IEEE Transactions on Fuzzy Systems, 22*(5), 1162–1182.
14. Mendel, J. M., Liu, F., & Zhai, D. (2009). σ-plane representation for type-2 fuzzy sets: Theory and applications. *IEEE Transactions on Fuzzy Systems, 17*(5), 1189–1207.
15. Mendel, J. M., John, R. I., & Liu, F. (2006). Interval type-2 fuzzy logic systems made simple. *IEEE transactions on Fuzzy Systems, 14*(6), 808–821.
16. Mersadek, M., Nagi, F., Permal, N., Ramasamy, A. A., & Azwin, A. (2024). Takagi-sugeno type 1–2 fuzzy linear output controller for two-area load frequency control. *Systems and Soft Computing, 6*, Article 200083.
17. Moali, O., Mezghani, D., Mami, A., Oussar, A., & Nemra, A. (2024). UAV trajectory tracking using proportional-integral-derivative-type-2 fuzzy logic controller with genetic algorithm parameter tuning. *Sensors, 24*(20), 6678.
18. Nguyen, A. T., Dinh, T. Q., Guerra, T. M., & Pan, J. (2021). Takagi-Sugeno fuzzy unknown input observers to estimate nonlinear dynamics of autonomous ground vehicles: Theory and real-time verification. *IEEE/ASME Transactions on Mechatronics, 26*(3), 1328–1338.
19. Pulido, M., Melin, P., Castillo, O., & Castro, J. R. (2024). Comparison of interval type-3 mamdani and sugeno models for fuzzy aggregation applied to ensemble neural networks for Mexican stock exchange time series prediction. *Mathematical and Computational Applications, 29*(4), 67.
20. Tavoosi, J., Shirkhani, M., Abdali, A., Mohammadzadeh, A., Nazari, M., Mobayen, S., & Bartoszewicz, A. (2021). A new general type-2 fuzzy predictive scheme for PID tuning. *Applied Sciences, 11*(21), 10392.

21. Teodorović, D., Davidović, T., Šelmić, M., & Nikolić, M.: *Bee colony optimization and its applications. Handbook of AI-based Metaheuristics* (pp. 301–322).
22. Teodorovic, D., Lucic, P., Markovic, G., & Dell'Orco, M. (2006). Bee colony optimization: principles and applications. In *2006 8th seminar on neural network applications in electrical engineering* (pp. 151–156). IEEE.
23. Top, A., & Gökbulut, M. (2024). Design of an Android application for autonomous control of mobile robot. *Niğde Ömer Halisdemir Üniversitesi Mühendislik Bilimleri Dergisi, 13*(2), 448–459.
24. Vignesh, C., Uma, M., & Sethuramalingam, P. (2024). Development of rapidly exploring random tree based autonomous mobile robot navigation and velocity predictions using K-nearest neighbors with fuzzy logic analysis. *International Journal on Interactive Design and Manufacturing (IJIDeM)*, 1–25.
25. Wang, B., Rong, N., & Ji, Y. (2024). Event-based predictive control for interval type-2 Takagi-Sugeno fuzzy systems under Kalman filter. *Proceedings of the Institution of Mechanical Engineers, Part I: Journal of Systems and Control Engineering, 238*(5), 848–859.
26. Zadeh, L. A. (1975). The concept of a linguistic variable and its application to approximate reasoning. *Part II, Information Sciences, 8*, 301–357.
27. Zhao, T., Chen, Y., Dian, S., Guo, R., & Li, S. (2020). General type-2 fuzzy gain scheduling PID controller with application to power-line inspection robots. *International Journal of Fuzzy Systems, 22*(1), 181–200.
28. Zhou, J., & Chen, Y. (2024). Study on non-iterative algorithms for center-of-sets type-reduction of interval type-2 takagi–sugeno–kang fuzzy logic systems. *International Journal of Fuzzy Systems*, 1–13.
29. Amador-Angulo, L., Castillo, O., Melin, P., & Geem, Z. G. (2024). Type-3 fuzzy dynamic adaptation of Bee colony optimization applied to mathematical functions. *Fuzzy Sets and Systems, 489*. https://doi.org/10.1016/j.fss.2024.109014

Type-3 Fuzzy Modeling in the Decision of Process for House Purchasing

Patricia Melin

Abstract This paper outlines a novel approach for modeling the decision-process of house purchasing utilizing type-3 fuzzy logic. The idea is representing the uncertainty in the decision-making process with type-3 fuzzy logic, which can surpass type-2 and type-1 in this task. The type-3 fuzzy model consists of fuzzy rules that contain expert knowledge in the decision-making process, while type-3 membership functions represent the uncertainty of the linguistic concepts involved in obtaining a decision on house buying. The type-3 fuzzy model was tested with an illustrative application to validate the correctness of the proposal and in the future, we will consider comparing with type-2 and type-1 models to show the advantages of type-3.

Keywords Type-3 fuzzy logic · Decision-making · House purchasing

1 Introduction

Fuzzy logic since its inception [1] has been successfully utilized in a plethora of areas [2, 3]. The main idea of fuzzy logic (FL) was to model linguistic uncertainty in real human decision-making, and the original FL is now called type-1 as it was the initial version of this idea. More recently, type-2 was proposed and developed to offer a better model of uncertainty [4] and consequently more complex problems were solved [5, 6]. Still, more complex problems can be tackled with a most recent version of fuzzy theory, called type-3 fuzzy logic [7–9], which offers an even better model of uncertainty a now some recent successful applications have been achieved, like in [10–17].

For the particular case of car selection or purchasing, several previous works can be found, but mostly with type-1 fuzzy logic [18–21]. These previous works consider this problem from different perspectives, but in general the decision-making process is viewed under uncertainty. In this sense, we have identified a research gap in this area,

P. Melin (✉)
Tijuana Institute of Technology, TecNM, Tijuana, Mexico
e-mail: pmelin@tectijuana.mx

O. Castillo and P. Melin (eds.), *Intelligent System Design Based on Soft Computing Models*, Studies in Computational Intelligence 1223,
https://doi.org/10.1007/978-3-031-97309-3_4

which we believe is worth doing some research work on it. For this reason, we decided to consider modeling the decision-making problem of car selection or purchasing by utilizing type-3 fuzzy logic. Accordingly, the contribution is the proposed type-3 fuzzy approach for modeling intention of purchasing a house of customers. The aim of type-3 is to capture in a better way the involved uncertainty in decision-making. We have to say that this is the first work on the utilization of type-3 fuzzy modeling for this decision-making process, although is just a first contribution in this area and there are many improvements that could be done in the future.

The paper is organized as follows. In Sect. 2 we offer some concepts of type-3 FL. Section 3 describes the type-3 fuzzy model for house buying. Section 4 summarizes the illustrative application to verify the proposal, and Sect. 5 offers the conclusions.

2 Type-3 Fuzzy Logic

This Section reviews some basic concepts about type-3 that are needed for the reader to understand the idea of the model in the following section. The most basic change starts from the concept of a type-3 fuzzy set.

In type-3 FL, we could handle higher degrees of uncertainty with respect to type-2 due to the topology of the membership functions (MFs). A Type-3 fuzzy set (T3 FS) [7] represented by $A^{(3)}$, is the graph of a trivariate function, named MF of $A^{(3)}$, in the Cartesian product expressed by Eq. (1), where the primary variable of $A^{(3)}$ has a universe X, x. The MF of $\mu_{A^{(3)}}$ is formulated by $\mu_{A^{(3)}}(x, u, v)$, and is a Type-3 MF of the T3 fuzzy set postulated by Eq. (2):

$$\mu_{A^{(3)}} : X \times [0,1] \times [0,1] \rightarrow [0,1] \tag{1}$$

$$\begin{aligned} &A(x, u(x), v(x, u), \mu_{A^{(3)}}(x, u, v))^{(3)} \\ &= \{|x \in X, u \in U \subseteq [0, 1], v \in V \subseteq [0, 1]\} \end{aligned} \tag{2}$$

where u is the secondary variable and has the universe U, and V for tertiary variable v. Based on (1) and (2), the concepts of type-3 fuzzy relations, inference, logic and fuzzy systems can be postulated in an analogous form to type-2 (more details can be found in [15]).

3 Type-3 for House Purchasing

In this Section, the type-3 fuzzy model is outlined in more detail. First, we define the linguistic variables and values of the fuzzy system, followed by the rules and MFs.

Table 1 Information about the inputs of the fuzzy model

Nomenclature	Linguistic variable	Linguistic values	Ranges
P	Price	(Cheap, normal, expensive)	[0, 4]
Sz	Size	(Small, regular, big)	[0, 4]
Sa	Safety	(Bad, medium, good)	[0, 4]
Q	Quality	(Bad, normal, excellent)	[0, 4]

Table 2 Information about the output of the fuzzy model

Nomenclature	Linguistic values	Ranges
H (standard house)	H_1, H_2, H_3	[0, 10]
R (private residence)	R_1, R_2, R_3	[0, 10]
C (condominium)	C_1, C_2, C_3	[0, 10]

Based on the expert knowledge available for this problem, we were able to identify the relevant input variables that can determine the selection of a particular house to be purchased.

In Table 1 we show the information about the inputs of the type-3 system, including the nomenclature, the names of the linguistic variables and values, and corresponding ranges.

We are assuming, in this case, that the real ranges of price and size of the house are mapped to the range of [0, 4] for the ease of implementation. In addition, the particular granulation of the linguistic variables was done according to the experts in the area.

In Table 2 we show the information about the output variable of the type-3 system, which is a selected house (SH) to purchase. We are assuming that there are 3 possible types of standard house, 3 possible types of private residence, and 3 types of condominiums. The increasing order in index denotes higher quality or luxury of the house.

In Table 3 we show the fuzzy rules containing the expert knowledge about the decision-making involved in the house purchasing process.

4 Illustrative Example

This Section summarizes an illustrative example of house purchasing and the results achieved with the type-3 model for a set of particular input values. We will assume that the type-3 fuzzy sets are scale Gaussian numbers with parameters $(\mu, \sigma, \lambda, \ell)$, which are the center, deviation, scale and lag parameters [7]. In this way, we would have the following fuzzy numbers for the values of the fuzzy system outputs:

Table 3 Fuzzy rules for house selection

P	Sz	Sa	Q	SH
Expensive	Big	Good	Excellent	R_3
Expensive	Regular	Good	Excellent	C_3
Expensive	Big	Medium	Excellent	H_3
Expensive	Big	Good	Normal	R_3
Expensive	Regular	Good	Normal	C_2
Expensive	Big	Medium	Normal	H_2
Normal	Small	Good	Excellent	C_1
Normal	Regular	Medium	Normal	H_2
Normal	Regular	Good	Bad	R_2
Normal	Big	Bad	Bad	H_1
Cheap	Regular	Bad	Bad	H_1
Cheap	Small	Medium	Bad	R_1
Cheap	Big	Bad	Bad	H_1
Cheap	Small	Medium	Normal	R_1

$$
\begin{aligned}
H_1 &= \text{scale Gausss scale Gauss}(1,\ 0.3,\ 0.9,\ 0.1)\\
H_2 &= \text{scaleGausssscaleGauss}(2,\ 0.3,\ 0.9,\ 0.1)\\
H_3 &= \text{scaleGausssscaleGauss}(3,\ 0.3,\ 0.9,\ 0.1)\\
R_1 &= \text{scaleGausssscaleGauss}(4,\ 0.3,\ 0.9,\ 0.1)\\
R_2 &= \text{scaleGausssscaleGauss}(5,\ 0.3,\ 0.9,\ 0.1)\\
R_3 &= \text{scaleGausssscaleGauss}(6,\ 0.3,\ 0.9,\ 0.1)\\
C_1 &= \text{scaleGausssscaleGauss}(7,\ 0.3,\ 0.9,\ 0.2)\\
C_2 &= \text{scaleGausssscaleGauss}(8,\ 0.3,\ 0.9,\ 0.2)\\
C_3 &= \text{scaleGausssscaleGauss}(9,\ 0.3,\ 0.9,\ 0.2)
\end{aligned}
$$

The fuzzy system inputs can be defined as follows. First, for the P input:

$$
\begin{aligned}
\text{cheap} &= \text{scaleGausssscaleGauss}(1,\ 0.3,\ 0.9,\ 0.1)\\
\text{normal} &= \text{scaleGausssscaleGauss}(2,\ 0.3,\ 0.9,\ 0.1)\\
\text{expensive} &= \text{scaleGausssscaleGauss}(3,\ 0.3,\ 0.9,\ 0.1)
\end{aligned}
$$

Now for the second input (Sz) we have the following:

$$
\begin{aligned}
\text{small} &= \text{scaleGausssscaleGauss}(1,\ 0.3,\ 0.9,\ 0.1)\\
\text{regular} &= \text{scaleGausssscaleGauss}(2,\ 0.3,\ 0.9,\ 0.1)\\
\text{big} &= \text{scaleGausssscaleGauss}(3,\ 0.3,\ 0.9,\ 0.1
\end{aligned}
$$

Now for the third input (Sa) we have the linguistic values:

$$\text{bad} = \text{scaleGausssscaleGauss}(1, 0.3, 0.9, 0.1)$$
$$\text{medium} = \text{scaleGausssscaleGauss}(2, 0.3, 0.9, 0.1)$$
$$\text{good} = \text{scaleGausssscaleGauss}(3, 0.3, 0.9, 0.1)$$

The fuzzy sets for the fourth input (Q) are as follows:

$$\text{bad} = \text{scaleGausssscaleGauss}(1, 0.3, 0.9, 0.1)$$
$$\text{normal} = \text{scaleGausssscaleGauss}(2, 0.3, 0.9, 0.1)$$
$$\text{excellent} = \text{scaleGausssscaleGauss}(3, 0.3, 0.9, 0.1)$$

Figure 1 shows the MFs for the price input variable. Figure 2 exhibits the MFs of the size input variable. Figures 3 and 4 depicts the Safety variable, while Figure shows the quality variable. Lastly, Fig. 5 shows the output variable.

We also show in Fig. 6 a 3D representation of the output variable, just as an illustration.

Next, we consider a set of input numeric values to verify the correctness of the model:

From Table 4 we can notice that all the results are as we expected, which means in all cases the result is the same as the experts with this we are able to verify the correctness of the model. We still need to consider other input values, as well as compare results with respect to type-2 and type-1 versions of the system.

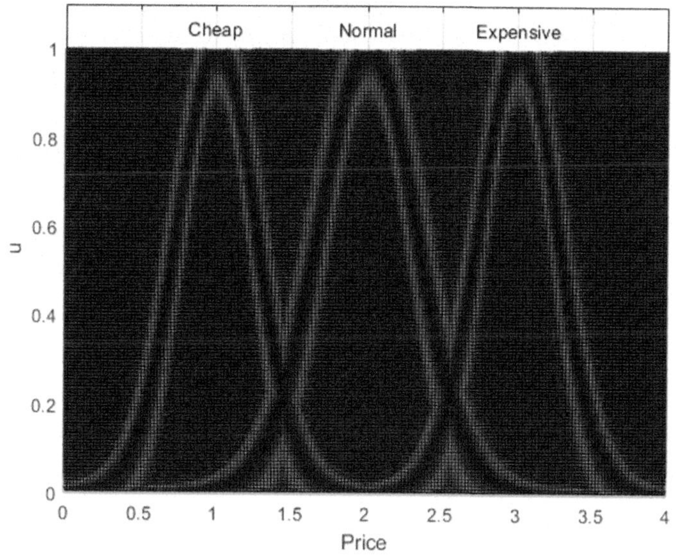

Fig. 1 MFs of the price variable

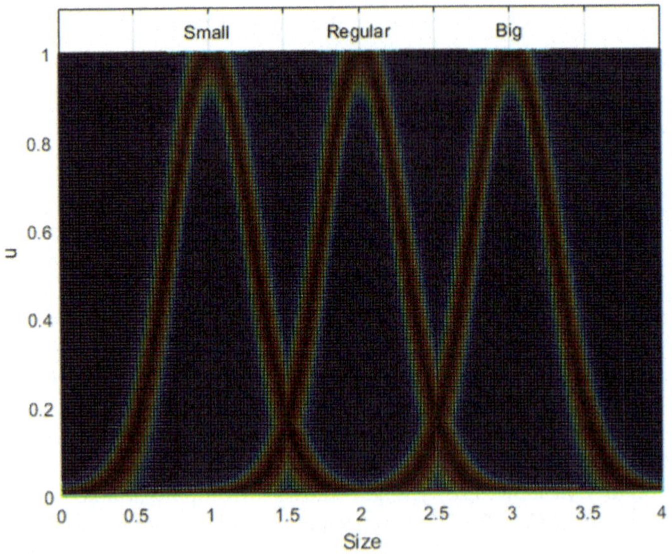

Fig. 2 MFs of the size variable

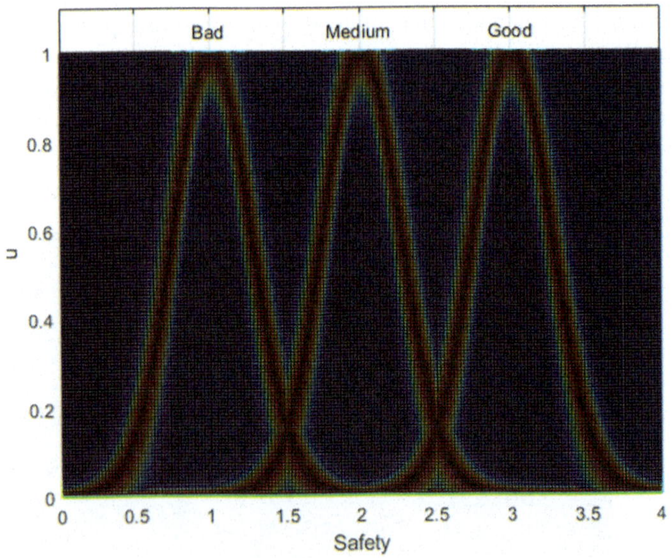

Fig. 3 MFs of the safety variable

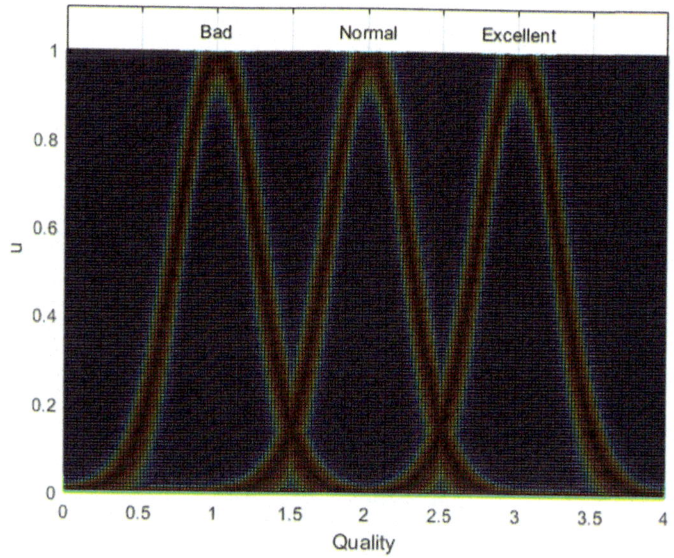

Fig. 4 MFs of the quality variable

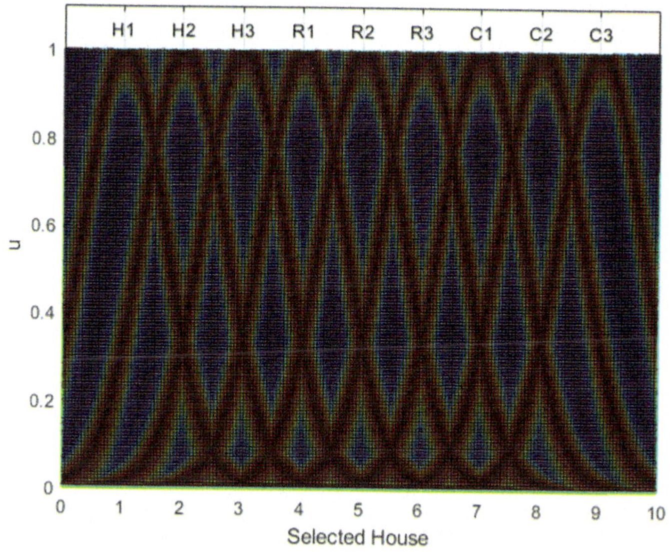

Fig. 5 MFs of the output variable

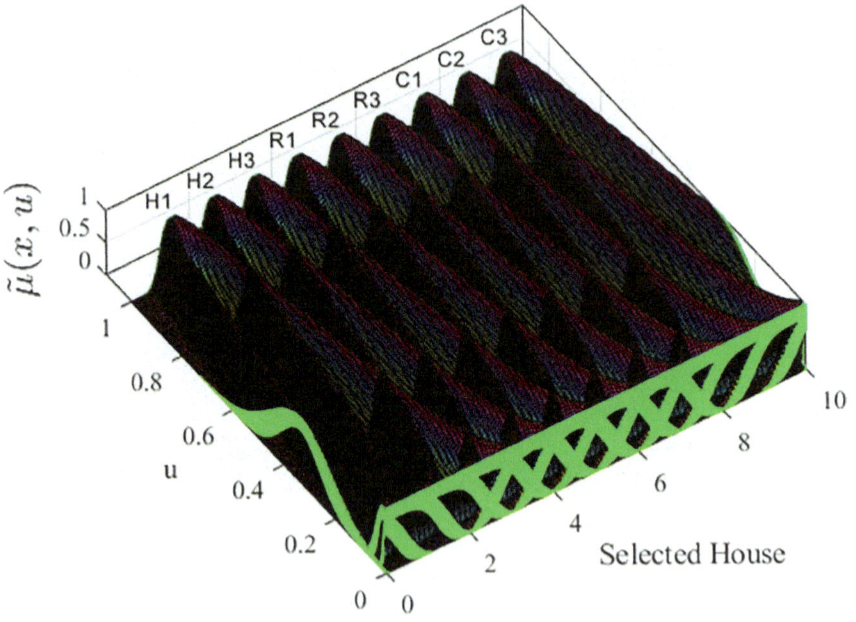

Fig. 6 MFs of the output in 3D

Table 4 Results for different input values in house selection

P	Sz	Sa	Q	SH
3	3	3	3	6 (R_3)
3	2	3	3	9 (C_3)
3	3	2	3	3 (H_3)
3	3	3	2	6 (R_3)
3	2	3	2	8 (C_2)
3	3	2	2	2 (H_2)
2	1	3	3	7 (C_1)
2	2	2	2	2 (H_2)
2	2	3	1	5 (R_2)
2	3	1	1	1 (H_1)
1	2	1	1	1 (H_1)
1	1	2	1	4 (R_1)
1	3	1	1	1 (H_1)
1	1	2	2	4 (R_1)

5 Conclusions

This paper described an approach for modeling the decision-process of house purchasing utilizing type-3 FL. The idea of the approach was representing the uncertainty in the house purchasing decision-making process with type-3 FL. The type-3 fuzzy model consisted of fuzzy rules that contained expert knowledge in the decision-making process, while type-3 membership functions represented the uncertainty of the linguistic concepts involved in decision-making. The type-3 fuzzy model was tested with an illustrative application to verify the correctness of the same. As future work, we plan to compare the type-3 proposal with respect to type-2 and type-1 models to show the potential advantages of type-3, in this way showing that type-3 has the capability of surpassing the other fuzzy models. In addition, the optimization of parameters in the fuzzy model is also a possible future work, like in [22–26].

References

1. Zadeh, L. A. (1965). Fuzzy sets. *Information and Control, 8*, 338–353.
2. Zadeh, L. A. (1989). Knowledge representation in fuzzy logic. *IEEE Transactions on Knowledge Data Engineering, 1*, 89.
3. Zadeh, L. A. (1975). The concept of a linguistic variable and its application to approximate reasoning. *Information Sciences, 8*, 199–249.
4. Mendel, J. M., & Bob John, R. I. (2002). Type-2 fuzzy sets made simple. *IEEE Transactions on Fuzzy Systems, 10*(2), 117–127.
5. Ontiveros, E., Melin, P., & Castillo, O. (2020). Comparative study of interval Type-2 and general Type-2 fuzzy systems in medical diagnosis. *Information Sciences, 525*, 37–53.
6. Ontiveros-Robles, E., Castillo, O., & Melin, P. (2021). Towards asymmetric uncertainty modeling in designing general type-2 fuzzy classifiers for medical diagnosis. *Expert Systems with Applications, 183*, Article 115370.
7. Castillo, O., & Melin, P. (2022). Towards interval type-3 intuitionistic fuzzy sets and systems. *Mathematics, 10*, 4091. https://doi.org/10.3390/math10214091
8. Qasem, S. N., Ahmadian, A., Mohammadzadeh, A., Rathinasamy, S., & Pahlevanzadeh, B. (2021). A type-3 logic fuzzy system: Optimized by a correntropy based Kalman filter with adaptive fuzzy kernel size. *Information Sciences, 572*, 424–443.
9. Mohammadzadeh, A., Sabzalian, M. H., & Zhang, W. (2020). An interval type-3 fuzzy system and a new online fractional-order learning algorithm: Theory and practice. *IEEE Transactions on Fuzzy Systems, 28*(9), 1940–1950.
10. Liu, Z., Mohammadzadeh, A., Turabieh, H., Mafarja, M., Band, S. S., & Mosavi, A. (2021). A new online learned interval type-3 fuzzy control system for solar energy management systems. *IEEE Access, 9*, 10498–10508.
11. Singh, D., Verma, N. K., Ghosh, A. K., & Malagaudanavar, A. K. (2021). An approach towards the design of interval type-3 TS fuzzy system. *IEEE Transactions on Fuzzy Systems.*
12. Wang, J. H., Tavoosi, J., Mohammadzadeh, A., Mobayen, S., Asad, J. H., Assawinchaichote, W., & Skruch, P. (2021). Non-singleton type-3 fuzzy approach for flowmeter fault detection: Experimental study in a gas industry. *Sensors, 21*(21), 7419.
13. Alattas, K. A., Mohammadzadeh, A., Mobayen, S., Aly, A. A., & Felemban, B. F. (2021). A new data-driven control system for MEMSs gyroscopes: dynamics estimation by type-3 fuzzy systems. *Micromachines, 12*(11), 1390.

14. Cao, Y., Raise, A., Mohammadzadeh, A., Rathinasamy, S., Band, S. S., & Mosavi, A. (2021). Deep learned recurrent type-3 fuzzy system: Application for renewable energy modeling/prediction. *Energy Reports, 7*, 8115–8127.
15. Amador-Angulo, L., Castillo, O., Castro, J., & Melin, P. (2023). A new approach for interval type-3 fuzzy control of nonlinear plants. *International Journal of Fuzzy Systems, 25*, 1624–1642.
16. Castillo, O., Valdez, F., Melin, P., & Ding, W. (2024). A survey on type-3 fuzzy logic systems and their control applications. *IEEE/CAA Journal of Automatica Sinica, 11*(8), 1744–1756. https://doi.org/10.1109/JAS.2024.124530
17. Ochoa, P., Peraza, C., Melin, P., Castillo, O., Park, S., & Geem, Z. W. (2024). Enhancing control systems through type-3 fuzzy logic optimization. *Mathematics, 12*, 1792. https://doi.org/10.3390/math12121792
18. Masui, S., Terano, T., Yumino, M., & Mimori, S. (1994). Decision making support system for house purchase by using fuzzy logic. *Computers & Industrial Engineering, 27*(1–4), 281–284.
19. Lopez, V., Montero, J., & del Monte, A. (2008). Fuzzy logic in real estate valuation. *Computational Intelligence in Decision and Control, World Scientific, 1*, 1021–1026.
20. Lopez, V., Santos, M., & Montero, J. (2010). Fuzzy specification in real estate market decision making. *International Journal of Computational Intelligence Systems, 3*(1), 8–20.
21. Del Giudice, V., De Paola, P., & Cantisani, G. B. (2017). Valuation of real estate investments through fuzzy logic. *Buildings, 7*, 26. https://doi.org/10.3390/buildings7010026
22. Amador-Angulo, L., Castillo, O., Melin, P., & Geem, Z. G. (2024). Type-3 fuzzy dynamic adaptation of Bee colony optimization applied to mathematical functions. *Fuzzy Sets and Systems, 489*. https://doi.org/10.1016/j.fss.2024.109014
23. Santiago, A., Dorronsoro, B., Nebro, A. J., Durillo, J. J., Castillo, O., & Fraire, H. J. (2019). A novel multi-objective evolutionary algorithm with fuzzy logic based adaptive selection of operators: FAME. *Information Sciences, 471*, 233–251.
24. Martínez-Soto, R., Castillo, O., & Aguilar, L. T. (2014). Type-1 and Type-2 fuzzy logic controller design using a hybrid PSO–GA optimization method. *Information Sciences, 285*, 35–49.
25. González, B., Valdez, F., Melin, P., & Prado-Arechiga, G. (2015). Fuzzy logic in the gravitational search algorithm for the optimization of modular neural networks in pattern recognition. *Expert Systems with Applications, 42*(14), 5839–5847.
26. Castillo, O., Melin, P., Ontiveros, E., Peraza, C., Ochoa, P., Valdez, F., & Soria, J. (2019). A high-speed interval type 2 fuzzy system approach for dynamic parameter adaptation in metaheuristics. *Engineering Applications of Artificial Intelligence, 85*, 666–680. https://doi.org/10.1016/j.engappai.2019.07.020

Type-3 Fuzzy Modeling of Restaurant Selection Decision-Making

Oscar Castillo

Abstract This paper outlines a novel approach for modeling the decision-process of selecting a restaurant utilizing type-3 fuzzy logic (FL). The key idea of the approach is representing the uncertainty in decision-making with type-3 FL, which in theory can surpass type-2 and type-1 in this task. The type-3 fuzzy model consists of fuzzy rules that contain expert knowledge in the decision-making process, while type-3 membership functions represent the uncertainty of the linguistic concepts involved in obtaining a decision. The type-3 fuzzy model was tested with an illustrative application to verify its correctness. In the future, we also plan to compare the results of type-3 with the type-2 and type-1 models to highlight the advantages of the proposal.

Keywords Type-3 fuzzy logic · Decision-making · Restaurant selection

1 Introduction

Fuzzy logic since its inception [1] has been successfully utilized in a plethora of areas [2, 3]. The main idea of fuzzy logic (FL) was to model linguistic uncertainty in real human decision-making, and the original FL is now called type-1 as it was the initial version of this idea. More recently, type-2 was proposed and developed to offer a better model of uncertainty [4] and consequently more complex problems were solved [5, 6]. Still, more complex problems can be tackled with a most recent version of fuzzy theory, called type-3 fuzzy logic [7–9], which offers an even better model of uncertainty a now some recent successful applications have been achieved, like in [10–17].

For the particular case of restaurant selection, several previous works can be found, but mostly with type-1 fuzzy logic [18–21]. These previous works consider this problem from different perspectives, but in general the decision-making process is viewed as a problem under uncertainty. In this sense, we have identified a research

O. Castillo (✉)
Tijuana Institute of Technology, TecNM, Tijuana, Mexico
e-mail: ocastillo@tectijuana.mx

O. Castillo and P. Melin (eds.), *Intelligent System Design Based on Soft Computing Models*, Studies in Computational Intelligence 1223,
https://doi.org/10.1007/978-3-031-97309-3_5

gap in this area, which we believe is worth doing some research work on it. For this reason, we decided to consider modeling the decision-making problem of restaurant selection or purchasing by utilizing type-3 FL. Accordingly, the contribution is the proposed type-3 fuzzy approach for modeling restaurant selection by the customers. The aim of type-3 is to capture in a better fashion the intrinsic uncertainty in decision-making. We have to say that this is the first work on employing type-3 fuzzy modeling for this decision-making process, although is just a first contribution in this area and there are many improvements that could be done in the future.

The paper is organized as follows. In Sect. 2 we offer some terminology of type-3 FL. Section 3 describes the type-3 fuzzy model for restaurant selection. Section 4 summarizes simulation results, and Sect. 5 offers the conclusions.

2 Type-3 Fuzzy Logic

This Section reviews some concepts about type-3 that are needed for the reader to understand the idea of the model. The most basic change starts from the concept of a type-3 fuzzy set.

Type-3 fuzzy logic can handle higher degrees of uncertainty with respect to type-2 due to the topology of the membership functions (MFs). A Type-3 fuzzy set (T3 FS) [7] represented by $A^{(3)}$, is the graph of a trivariate function, named MF of $A^{(3)}$, in the Cartesian product defined by Eq. (1), where the primary variable of $A^{(3)}$ has a universe X, x. The MF of $\mu_{A^{(3)}}$ is formulated by $\mu_{A^{(3)}}(x, u, v)$, and is a Type-3 MF of the T3 fuzzy set postulated by Eq. (2):

$$\mu_{A^{(3)}} : X \times [0, 1] \times [0, 1] \rightarrow [0, 1] \tag{1}$$

$$\begin{aligned} A^{(3)} = \{&(x, u(x), v(x, u),\ \mu_{A^{(3)}}(x, u, v)) \\ &| x \in X,\ u \in U \subseteq [0, 1], v \in V \subseteq [0, 1]\} \end{aligned} \tag{2}$$

where u is the secondary variable and has the universe U, and V for tertiary variable v. Based on (1) and (2), the concepts of type-3 fuzzy relations, inference, logic and fuzzy systems can be postulated in an analogous form to type-2 (more details in [15]).

3 Type-3 for Restaurant Selection

In this Section, the type-3 fuzzy model is outlined in more detail. First, we define the linguistic variables and values of the fuzzy system, followed by the rules and MFs.

Based on the knowledge available for this problem, we were able to identify the relevant input variables that can determine the selection of a particular restaurant.

In Table 1 we show the information about the inputs of the type-3 fuzzy system, including the nomenclature, the names of the linguistic variables and values, and corresponding ranges. We are assuming, in this case, that the real range of budget is mapped to the range of [0, 4] for the ease of implementation. In addition, the particular granulation of the linguistic variables was done according to the experts in the area.

The output variable of the system is the preference level (PL) for a restaurant. We are assuming that there are 5 possible preference levels: very low VL), low (L), normal (N), high (H), and very high (VH). Ranges are assumed to be in the [0, 6] interval.

In Table 2 we show the rules containing the expert knowledge about the decision-making involved in the restaurant selection process.

Table 1 Information about the inputs of the fuzzy model

Nomenclature	Linguistic variable	Linguistic values	Ranges
T	Taste	(low, medium, high)	[0, 4]
B	Budget	(low, medium, high)	[0, 4]
L	Location	(near, medium, far)	[0, 4]
E	Environment	(informal, casual, formal)	[0, 4]
UR	User Review	(low, medium, high)	[0, 4]

Table 2 Fuzzy rules for restaurant selection

T	B	L	E	UR	PL
Low	High	Far	Casual	Low	VL
High	Low	Far	Casual	Low	L
Medium	Low	Far	Casual	Low	VL
Low	High	Near	Formal	High	VL
Medium	Medium	Far	Casual	Medium	VL
High	High	Far	Casual	Medium	N
Medium	Medium	Far	Casual	Medium	L
Medium	Medium	Medium	Casual	Medium	N
Medium	Medium	Medium	Informal	Medium	N
High	Medium	Medium	Formal	High	H
High	Medium	Near	Casual	High	H
Medium	Medium	Near	Casual	Medium	H
High	High	Near	Formal	High	VH
High	High	Near	Casual	High	VH

4 Illustrative Application

This Section summarizes an illustrative example and the results achieved with the type-3 model for a set of particular input values. We will assume that the type-3 fuzzy sets are scaled Gaussian numbers with parameters (μ, σ, λ, ℓ), which are the center, deviation, scale and lag parameters [7]. In this way, we would have the following fuzzy numbers for the values of the fuzzy system outputs:

$$\begin{aligned}
\mathrm{VL} &= \mathrm{scaleGausssscaleGauss}(1,\ 0.3,\ 0.9,\ 0.1)\\
\mathrm{L} &= \mathrm{scaleGausssscaleGauss}(2,\ 0.3,\ 0.9,\ 0.1)\\
\mathrm{N} &= \mathrm{scaleGausssscaleGauss}(3,\ 0.3,\ 0.9,\ 0.1)\\
\mathrm{H} &= \mathrm{scaleGausssscaleGauss}(4,\ 0.3,\ 0.9,\ 0.1)\\
\mathrm{VH} &= \mathrm{scaleGausssscaleGauss}(5,\ 0.3,\ 0.9,\ 0.1)
\end{aligned}$$

The fuzzy system inputs can be defined as follows. First, for the T input:

$$\begin{aligned}
\mathrm{low} &= \mathrm{scaleGausssscaleGauss}(1,\ 0.3,\ 0.9,\ 0.1)\\
\mathrm{medium} &= \mathrm{scaleGausssscaleGauss}(2,\ 0.3,\ 0.9,\ 0.1)\\
\mathrm{high} &= \mathrm{scaleGausssscaleGauss}(3,\ 0.3,\ 0.9,\ 0.1)
\end{aligned}$$

Now for the second input (B) we have the following:

$$\begin{aligned}
\mathrm{low} &= \mathrm{scaleGausssscaleGauss}(1,\ 0.3,\ 0.9,\ 0.1)\\
\mathrm{medium} &= \mathrm{scaleGausssscaleGauss}(2,\ 0.3,\ 0.9,\ 0.1)\\
\mathrm{high} &= \mathrm{scaleGausssscaleGauss}(3,\ 0.3,\ 0.9,\ 0.1)
\end{aligned}$$

Now for the third input (L) we have the linguistic values:

$$\begin{aligned}
\mathrm{near} &= \mathrm{scaleGausssscaleGauss}(1,\ 0.3,\ 0.9,\ 0.1)\\
\mathrm{medium} &= \mathrm{scaleGausssscaleGauss}(2,\ 0.3,\ 0.9,\ 0.1)\\
\mathrm{far} &= \mathrm{scaleGausssscaleGauss}(3,\ 0.3,\ 0.9,\ 0.1)
\end{aligned}$$

The fuzzy sets for the fourth input (E) are as follows:

$$\begin{aligned}
\mathrm{informal} &= \mathrm{scaleGausssscaleGauss}(1,\ 0.3,\ 0.9,\ 0.1)\\
\mathrm{casual} &= \mathrm{scaleGausssscaleGauss}(2,\ 0.3,\ 0.9,\ 0.1)\\
\mathrm{formal} &= \mathrm{scaleGausssscaleGauss}(3,\ 0.3,\ 0.9,\ 0.1)
\end{aligned}$$

The fuzzy sets for the fifth input (UR) are:

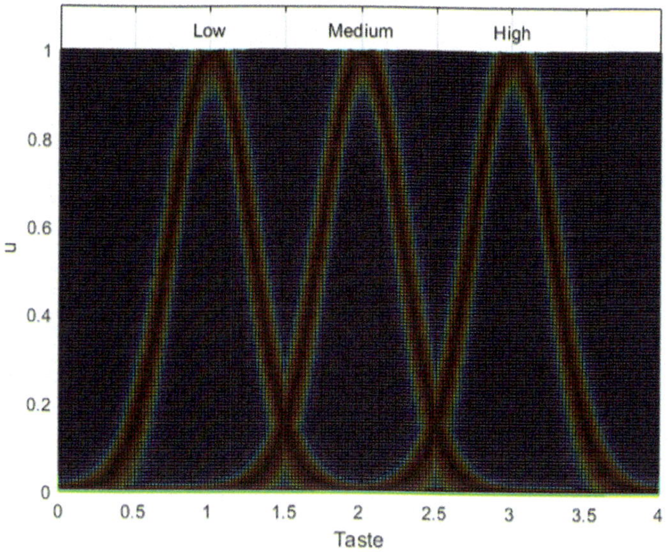

Fig. 1 MFs of the taste

$$\begin{aligned}\text{low} &= \text{scaleGausssscaleGauss}(1,\ 0.3,\ 0.9,\ 0.1)\\ \text{medium} &= \text{scaleGausssscaleGauss}(2,\ 0.3,\ 0.9,\ 0.1)\\ \text{high} &= \text{scaleGausssscaleGauss}(3,\ 0.3,\ 0.9,\ 0.1)\end{aligned}$$

Figure 1 depicts the MFs of the "taste" variable and Fig. 2 illustrates the MFs of the "budget" variable. Figure 3 exhibits the MFs of the "location" variable, Fig. 4 depicts the "environment" variable and Fig. 5 illustrates the "user review" variable. Lastly, Fig. 6 shows the output variable (preference level).

We also show, for illustration purposes, the 3D view of the MFs of the output variable (Fig. 7).

In Table 3 we consider a set of input numeric values to verify the correctness of the proposed model.

From Table 3 we can notice that all the results are as we expected, which means in all cases the result is the same as the experts with this we are able to verify the correctness of the model. We still need to consider other input values, as well as compare results with respect to type-2 and type-1 versions of the system.

5 Conclusions

This paper described an approach for modeling the decision-process of restaurant selection utilizing type-3 FL. The idea of the approach was representing the uncertainty in the decision-making process with type-3 FL. The type-3 fuzzy model

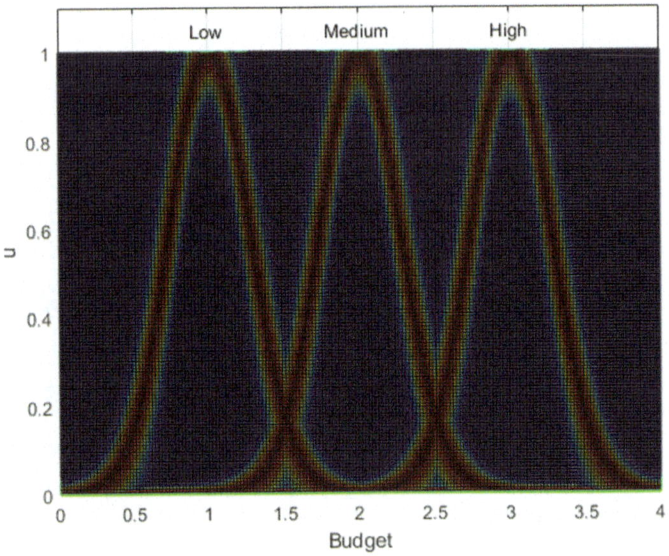

Fig. 2 MFs of the budget

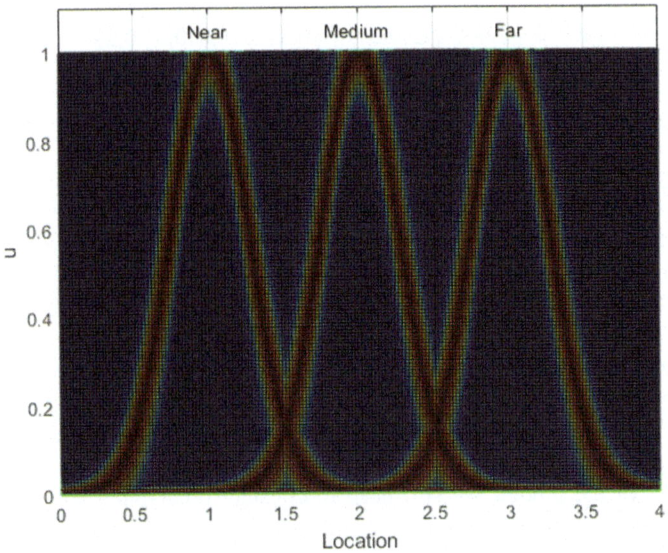

Fig. 3 MFs of the location

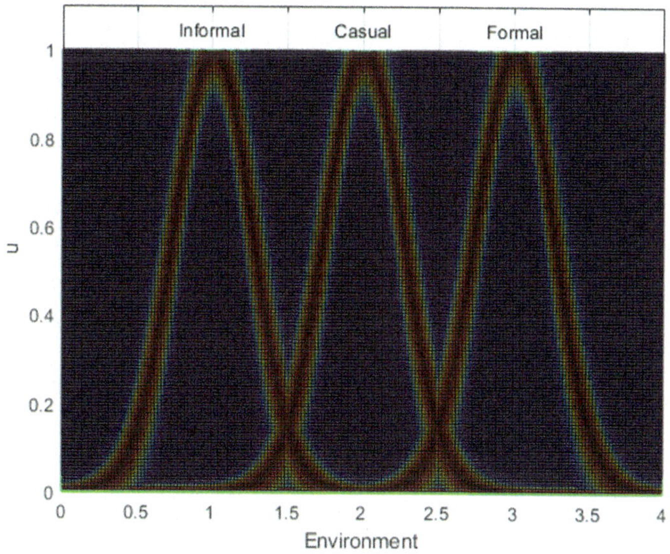

Fig. 4 MFs of the environment

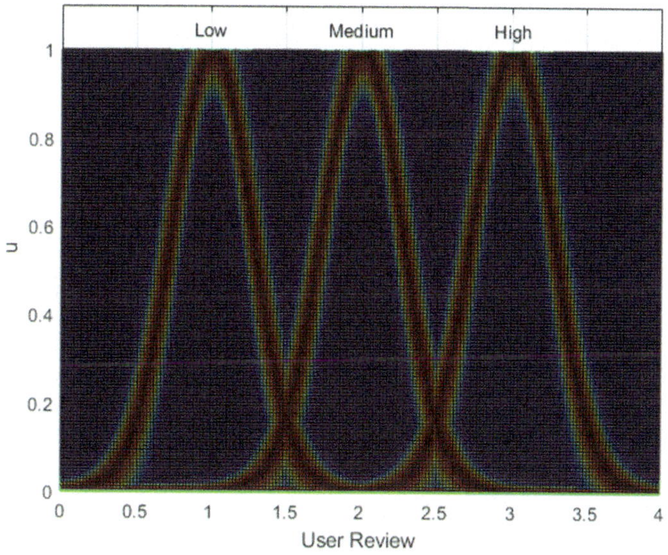

Fig. 5 MFs of the user review

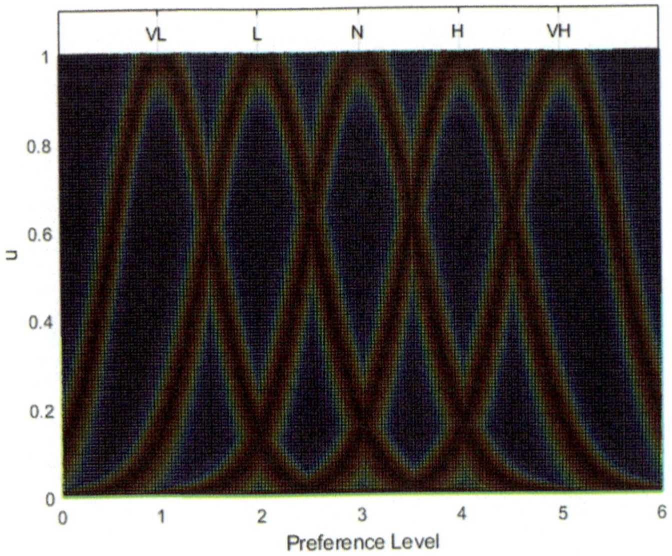

Fig. 6 MFs of the output variable

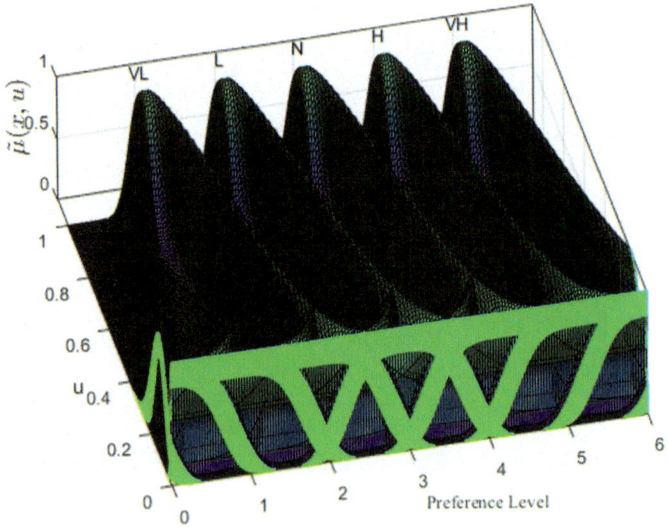

Fig. 7 MFs of the output variable in 3D view

consisted of fuzzy rules that contained expert knowledge in decision-making, while type-3 membership functions represented the uncertainty of the linguistic concepts involved in decision-making. The type-3 fuzzy model was tested with an illustrative application to verify its correctness. As future work, we are planning to compare the

Table 3 Simulation results for restaurant selection

T	B	L	E	UR	PL
1	3	3	2	1	1 (VL)
3	1	3	2	1	2 (L)
2	1	3	2	1	1 (VL)
1	3	1	3	3	1 (VL)
2	2	3	2	2	1 (VL)
3	3	3	2	2	3 (N)
2	2	3	2	2	2 (L)
2	2	2	2	2	3 (N)
2	2	2	1	2	3 (N)
3	2	2	3	3	4 (H)
3	2	1	2	3	4 (H)
2	2	1	2	2	4 (H)
3	3	1	3	3	5 (VH)
3	3	1	2	3	5 (VH)

results of type-3 with respect to the type-2 and type-1 models to show the advantages of the proposal, in this way showing that type-3 has the potential of surpassing the other fuzzy models. In addition, we could try the implementation in hardware of type-3 systems, as was done previously for type-2 [22]. Also, other application areas could be considered, such as control [23, 24] or pattern recognition [25].

References

1. Zadeh, L. A. (1965). Fuzzy sets. *Inf Control, 8*, 338–353.
2. Zadeh, L. A. (1989). Knowledge representation in Fuzzy Logic. *IEEE Transactions on Knowledge Data Engineering, 1*, 89.
3. Zadeh, L. A. (1975). The concept of a linguistic variable and its application to approximate reasoning. *Information Sciences, 8*, 199–249.
4. Mendel, J. M., & Bob John, R. I. (2002). Type-2 fuzzy sets made simple. *IEEE Transactions on Fuzzy Systems, 10*(2), 117–127.
5. Ontiveros, E., Melin, P., & Castillo, O. (2020). Comparative study of interval Type-2 and general Type-2 Fuzzy systems in medical diagnosis. *Information Sciences, 525*, 37–53.
6. Ontiveros-Robles, E., Castillo, O., & Melin, P. (2021). Towards asymmetric uncertainty modeling in designing General Type-2 Fuzzy classifiers for medical diagnosis. *Expert Systems with Applications, 183*, Article 115370.
7. Castillo, O., Castro, J. R., & Melin, P. (2022). *Interval type-3 fuzzy systems: theory and design*. Springer.
8. Qasem, S. N., Ahmadian, A., Mohammadzadeh, A., Rathinasamy, S., & Pahlevanzadeh, B. (2021). A type-3 logic fuzzy system: Optimized by a correntropy based Kalman filter with adaptive fuzzy kernel size. *Information Sciences, 572*, 424–443.

9. Mohammadzadeh, A., Sabzalian, M. H., & Zhang, W. (2020). An interval type-3 fuzzy system and a new online fractional-order learning algorithm: Theory and practice. *IEEE Transactions on Fuzzy Systems, 28*(9), 1940–1950.
10. Liu, Z., Mohammadzadeh, A., Turabieh, H., Mafarja, M., Band, S. S., & Mosavi, A. (2021). A new online learned interval type-3 fuzzy control system for solar energy management systems. *IEEE Access, 9*, 10498–10508.
11. Singh, D., Verma, N. K., Ghosh, A. K., & Malagaudanavar, A. K. (2021). An approach towards the design of interval type-3 ts fuzzy system. *IEEE Transactions on Fuzzy Systems.*
12. Wang, J. H., Tavoosi, J., Mohammadzadeh, A., Mobayen, S., Asad, J. H., Assawinchaichote, W., & Skruch, P. (2021). Non-singleton type-3 fuzzy approach for flowmeter fault detection: experimental study in a gas industry. *Sensors, 21*(21), 7419.
13. Alattas, K. A., Mohammadzadeh, A., Mobayen, S., Aly, A. A., & Felemban, B. F. (2021). A new data-driven control system for MEMSs gyroscopes: dynamics estimation by type-3 fuzzy systems. *Micromachines, 12*(11), 1390.
14. Cao, Y., Raise, A., Mohammadzadeh, A., Rathinasamy, S., Band, S. S., & Mosavi, A. (2021). Deep learned recurrent type-3 fuzzy system: Application for renewable energy modeling/prediction. *Energy Reports, 7*, 8115–8127.
15. Amador-Angulo, L., Castillo, O., Castro, J., & Melin, P. (2023). A new approach for interval type-3 fuzzy control of nonlinear plants. *International Journal of Fuzzy Systems, 25*, 1624–1642.
16. Castillo, O., Valdez, F., Melin, P., & Ding, W. (2024). A survey on type-3 fuzzy logic systems and their control applications. *IEEE/CAA Journal of Automatica Sinica, 11*(8), 1744–1756. https://doi.org/10.1109/JAS.2024.124530
17. Ochoa, P., Peraza, C., Melin, P., Castillo, O., Park, S., & Geem, Z. W. (2024). Enhancing control systems through type-3 fuzzy logic optimization. *Mathematics, 12*, 1792. https://doi.org/10.3390/math12121792
18. Stefano, N.M., Casarotto Filho, N., Barichello, R., & Sohn, A.P. (2015). A fuzzy SERVQUAL based method for evaluated of service quality in the hotel industry. *Procedia CIRP, 30*, pp. 433–438.
19. Djamal, N., & Widyarto, W. O. (2017). Quality services measurement using fuzzy service quality (fuzzy servqual) method. *Int J Eng Res Technol, 6*, 614–618.
20. Hartanto, M., & Nugeraha Utama, D. (2020). Intelligent decision support model for recommending restaurant. *Cogent Engineering, 7*(1), 1763888. https://doi.org/10.1080/23311916.2020.1763888
21. Masroor, I., Abrar, M., & Masroor, R. (2021). Customized food and restaurant expanding fuzzy logic. *The Journal of Mathematics and Computer Science, 11*, 2242–2262.
22. Sepúlveda, R., Montiel, O., Castillo, O., & Melin, P. (2012). Embedding a high speed interval type-2 fuzzy controller for a real plant into an FPGA. *Applied Soft Computing, 12*(3), 988–998.
23. Astudillo, L., Castillo, O., Melin, P., Alanis, A., Soria, J., & Aguilar, L.T. (2006). Intelligent control of an autonomous mobile robot using type-2 fuzzy logic. *Engineering Letters, 13*(3)
24. Mohammadzadeh, A., Castillo, O., Band, S. S., et al. (2021). A novel fractional-order multiple-model type-3 fuzzy control for nonlinear systems with unmodeled dynamics. *International Journal of Fuzzy Systems, 23*, 1633–1651. https://doi.org/10.1007/s40815-021-01058-1
25. Melin, P., Urias, J., Solano, D., Soto, M., Lopez, M., & Castillo, O. (2006). Voice recognition with neural networks, type-2 fuzzy logic and genetic algorithms. *Engineering Letters, 13*(3)

Application of InterCriteria Analysis on the Overall Household Expenditure by Expense Groups

Georgy Gilyov and Todor Petkov

Abstract In this paper the InterCriteria analysis is applied over the data of overall household expenditure by expense groups in Republic of Bulgaria. The data is provided from the National Statistical institute of Bulgaria and contains the information of the expense by different groups such as food expense, healthcare expense, transport expense etc. The two analyses were applied one by taking expense groups and one by taking years in order to show which of them corelate and how a household could optimize its finances.

Keywords InterCriteria analysis · Intuitionistic Fuzzy Pairs

1 Introduction

The concept of InterCriteria (ICA) Analysis is introduced in [3], it utilizes the theory of indexed matrices and intuitionistic fuzzy sets [1, 2, 4] and it gives the possibility to compare some criteria by their objects. During the years several extensions of the analysis were developed [5] along with various application in different fields such as medicine, industry, artificial intelligence and others [6–10].

The ICrA-method [3, 5] is based on two main concepts: intuitionistic fuzzy sets and index matrices. A brief description is given below.

Let I be a fixed set of indices and let R be the set of the real numbers. An index matrix (IM) with sets of indices K and L ($K, L \subset I$) is defined by (see [1, 4]).

G. Gilyov · T. Petkov (✉)
Department of Computer Systems and Technologies, "Prof. Asen Zlatarov" Univeristy–Burgas, Burgas, Bulgaria
e-mail: todor_petkov@btu.bg

G. Gilyov
e-mail: georgy009@abv.bg

O. Castillo and P. Melin (eds.), *Intelligent System Design Based on Soft Computing Models*, Studies in Computational Intelligence 1223,
https://doi.org/10.1007/978-3-031-97309-3_6

$$[K, L, \{a_{k_i,l_j}\}] \equiv \begin{array}{c|cccc} & l_1 & l_2 & \cdots & l_n \\ \hline k_1 & a_{k_1,l_1} & a_{k_1,l_2} & \cdots & a_{k_1,l_n} \\ k_2 & a_{k_2,l_1} & a_{k_2,l_2} & \cdots & a_{k_2,l_n} \\ \vdots & \vdots & \vdots & \ddots & \vdots \\ k_m & a_{k_m,l_1} & a_{k_m,l_2} & \cdots & a_{k_m,l_n} \end{array},$$

where $K = \{k_1, k_2, \ldots, k_m\}$, $L = \{l_1, l_2, \ldots, l_n\}$, for $1 \le i \le m$, and $1 \le j \le n$: $a_{k_i,l_j} \in R$.

For any two IMs, a series of relations, operations, and operators have been defined. The theory behind the IMs is described more thoroughly in [1].

Following the description of the ICA approach, given by [3, 5], let us consider an IM M with index sets with m rows $\{O_1, \ldots, O_m\}$ and n columns $\{C_1, \ldots, C_n\}$, where for every p, q ($1 \le p \le m$, $1 \le q \le n$), O_p in an evaluated object, C_q is an evaluation criterion, and e_{O_p,C_q} is the evaluation of the p-th object against the q-th criterion, defined as a real number or another object that is comparable according to relation R with all the rest elements of the IM M:

$$M = \begin{array}{c|ccccccc} & C_1 & \ldots & C_k & \ldots & C_l & \ldots & C_n \\ \hline O_1 & e_{O_1,C_1} & \ldots & e_{O_1,C_k} & \cdots & e_{O_1,C_l} & \ldots & e_{O_1,C_n} \\ \vdots & \vdots & \ddots & \vdots & \ddots & \vdots & \ddots & \vdots \\ O_i & e_{O_i,C_1} & \ldots & e_{O_i,C_k} & \cdots & e_{O_i,C_l} & \ldots & e_{O_i,C_n} \\ \vdots & \vdots & \ddots & \vdots & \ddots & \vdots & \ddots & \vdots \\ O_j & e_{O_j,C_1} & \ldots & e_{O_j,C_k} & \cdots & e_{O_j,C_l} & \cdots & e_{O_j,C_n} \\ \vdots & \vdots & \ddots & \vdots & \ddots & \vdots & \ddots & \vdots \\ O_m & e_{O_m,C_1} & \ldots & e_{O_m,C_j} & \cdots & e_{O_m,C_l} & \cdots & e_{O_m,C_n} \end{array},$$

From the requirement for comparability above, it follows that for each i, j, k it holds the relation $R(e_{O_i,C_k}, e_{O_j,C_k})$.

Let $S^{\mu}_{k,l}$ be the number of cases in which the relations $R(e_{O_i,C_k}, e_{O_j,C_k})$ and $R(e_{O_i,C_l}, e_{O_j,C_l})$ are simultaneously satisfied. Let also $S^{\nu}_{k,l}$ be the number of cases in which the relations $R(e_{O_i,C_k}, e_{O_j,C_k})$ and its dual $\overline{R} = (e_{O_i,C_l}, e_{O_j,C_l})$ are simultaneously satisfied. As the total number of pairwise comparisons between the objects is given by $m(m-1)/2$, it can be verified that the following inequalities hold:

$$0 \le S^{\mu}_{k,l} + S^{\nu}_{k,l} \le \frac{m(m-1)}{2}$$

.

For every k, l, such that $1 \le k \le l \le n$, and for $m \ge 2$ two numbers are defined:

$$\mu_{C_k,C_l} = 2\frac{S^{\mu}_{k,l}}{m(m-1)}, \ \nu_{C_k,C_l} = 2\frac{S^{\nu}_{k,l}}{m(m-1)}$$

2 Application of InterCriteria Analysis to Data on Overall Expenditure of Households in the Republic of Bulgaria.

InterCriteria Analysis has been applied to data on the overall expenditure of households within the Republic of Bulgaria. The information was sourced from the website of the National Statistical Institute [11]. The selected data are from the section "Demographic and Social Statistics.". The input data represents the different expenditure groups that a household spends money during the years from 2014 till 2023 the data for the ICA analysis is presented on the Table 1.

The initial analysis aimed to establish the relationships between various expense groups and the data were entered into the ICrAData software [9] with a selected option for expense analysis. Figure 1 displays the input data and the results obtained after applying the InterCriteria Analysis.

When applying the InterCriteria Analysis to expenditures, the following abbreviations were used:

- **A**—Total Consumer Expenditure
- **B**—Food
- **C**—Alcoholic Beverages and Tobacco Products
- **D**—Clothing and Footwear
- **E**—Housing, Water, Electricity, and Fuels
- **F**—Household Furnishing and Maintenance
- **G**—Healthcare
- **H**—Transport
- **I**—Communication
- **J**—Leisure, Cultural Recreation, and Education
- **K**—Miscellaneous Goods and Services
- **L**—Taxes
- **M**—Social Security Contributions
- **N**—Regular Transfers to Other Households
- **O**—Other Expenses
- **P**—Deposit
- **Q**—Paid Debts and Given Loans.

The results for the degrees of membership are shown on the Tables 2 and 3, the results for non-membership values are shown on the Tables 4 and 5. The overall result from the analysis is shown on the Table 6.

The analysis of the data reveals various levels of correlation between the expenditure criteria of households. We identify clearly expressed strong positive consonances between some categories, indicating a close connection between them.

- **Strong positive consonance** is primarily observed among categories related to basic life necessities such as food, healthcare, and housing expenses. This highlights that an increase in expenses in one of these categories often correlates with increases in the others.

Table 1 Input data for the overall expenditure of households within the Republic of Bulgaria

Year	2014	2015	2016	2017	2018	2019	2020	2021	2022	2023
Total consumer expenditure	9114	9328	9203	9835	10,493	11,037	10,890	12,107	14,358	16,451
Food	3534	3510	3432	3613	3834	3994	4034	4303	5210	5980
Alcoholic beverages and tobacco products	451	489	484	504	527	568	571	604	674	781
Clothing and footwear	416	396	402	421	438	440	404	536	580	696
Housing, water, electricity, and fuels	1468	1598	1595	1770	1760	1821	1855	2038	2445	2425
Household furnishing and maintenance	403	415	411	498	514	503	561	696	780	950
Healthcare	570	594	616	642	701	842	836	924	1126	1221
Transport	801	794	764	809	938	1017	893	1010	1266	1490
Communication	481	485	494	515	554	573	615	660	741	830
Leisure, cultural recreation, and education	511	545	530	528	623	702	539	653	783	1117
Miscellaneous goods and services	479	502	473	534	604	575	581	682	754	960
Taxes	531	555	579	666	712	759	775	825	872	1109
Social security contributions	688	712	727	822	939	1029	1078	1113	1230	1603
Regular transfers to other households	135	136	130	145	130	140	120	133	141	177
Other expenses	460	462	507	522	635	573	540	551	679	772
Deposit	274	315	457	648	682	836	1098	1116	1339	1394
Paid debts and given loans	440	402	365	354	300	299	292	296	291	405

- **Positive consonance** shows moderately strong links between different categories, which may include leisure and education, suggesting that these aspects of household life are connected to their overall financial situation.
- **Weak positive consonance and weak dissonance** identify weaker, yet still noticeable connections between categories, such as those between transportation expenses and other indirectly related categories.

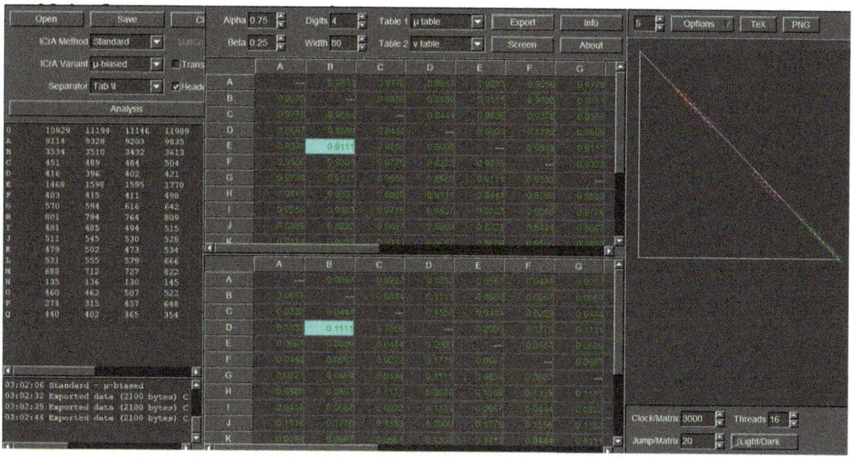

Fig. 1 Analysis of data on overall expenditure of households in the Republic of Bulgaria

Table 2 Results for membership values of overall expenditure of households in the Republic of Bulgaria (first part)

μ-table	A	B	C	D	E	F	G	H
A	1	0.9333	0.9778	0.8667	0.9333	0.9556	0.9778	0.9111
B	0.9333	1	0.9556	0.8889	0.9111	0.9333	0.9111	0.9333
C	0.9778	0.9556	1	0.8444	0.9556	0.9778	0.9556	0.8889
D	0.8667	0.8889	0.8444	1	0.8	0.8222	0.8889	0.9111
E	0.9333	0.9111	0.9556	0.8	1	0.9333	0.9111	0.8444
F	0.9556	0.9333	0.9778	0.8222	0.9333	1	0.9333	0.8667
G	0.9778	0.9111	0.9556	0.8889	0.9111	0.9333	1	0.8889
H	0.9111	0.9333	0.8889	0.9111	0.8444	0.8667	0.8889	1
I	0.9556	0.9333	0.9778	0.8667	0.9333	0.9556	0.9778	0.8667
J	0.8889	0.8222	0.8667	0.8	0.8222	0.8444	0.8667	0.8889
K	0.9111	0.9333	0.9333	0.8667	0.8889	0.9556	0.8889	0.9111
L	0.9556	0.9333	0.9778	0.8667	0.9333	0.9556	0.9778	0.8667
M	0.9556	0.9333	0.9778	0.8667	0.9333	0.9556	0.9778	0.8667
N	0.6444	0.6222	0.6222	0.6889	0.6222	0.6	0.6222	0.6889
O	0.8889	0.8222	0.8667	0.8444	0.8222	0.8889	0.9111	0.8889
P	0.9556	0.9333	0.9778	0.8667	0.9333	0.9556	0.9778	0.8667
Q	0.2444	0.2667	0.2222	0.3333	0.2222	0.2444	0.2222	0.3333

- **Dissonance** in its various degrees indicates that increases in expenditures in one category may not directly correspond or may even contradict trends in others. For example, regular transfers to other households show negative correlations with many other categories, which could suggest that higher transfers might reduce the ability to allocate funds in other areas.

Table 3 Results for membership values of overall expenditure of households in the Republic of Bulgaria (second part)

μ-table	I	J	K	L	M	N	O	P	Q
A	0.9556	0.8889	0.9111	0.9556	0.9556	0.6444	0.8889	0.9556	0.2444
B	0.9333	0.8222	0.9333	0.9333	0.9333	0.6222	0.8222	0.9333	0.2667
C	0.9778	0.8667	0.9333	0.9778	0.9778	0.6222	0.8667	0.9778	0.2222
D	0.8667	0.8	0.8667	0.8667	0.8667	0.6889	0.8444	0.8667	0.3333
E	0.9333	0.8222	0.8889	0.9333	0.9333	0.6222	0.8222	0.9333	0.2222
F	0.9556	0.8444	0.9556	0.9556	0.9556	0.6	0.8889	0.9556	0.2444
G	0.9778	0.8667	0.8889	0.9778	0.9778	0.6222	0.9111	0.9778	0.2222
H	0.8667	0.8889	0.9111	0.8667	0.8667	0.6889	0.8889	0.8667	0.3333
I	1	0.8444	0.9111	1	1	0.6	0.8889	1	0.2
J	0.8444	1	0.8444	0.8444	0.8444	0.6667	0.8667	0.8444	0.3556
K	0.9111	0.8444	1	0.9111	0.9111	0.6444	0.8889	0.9111	0.2889
L	1	0.8444	0.9111	1	1	0.6	0.8889	1	0.2
M	1	0.8444	0.9111	1	1	0.6	0.8889	1	0.2
N	0.6	0.6667	0.6444	0.6	0.6	1	0.6222	0.6	0.5778
O	0.8889	0.8667	0.8889	0.8889	0.8889	0.6222	1	0.8889	0.3111
P	1	0.8444	0.9111	1	1	0.6	0.8889	1	0.2
Q	0.2	0.3556	0.2889	0.2	0.2	0.5778	0.3111	0.2	1

Table 4 Results for non-membership values of overall expenditure of households in the Republic of Bulgaria (first part)

ν-table	A	B	C	D	E	F	G	H
A	0	0.0667	0.0222	0.1333	0.0667	0.0444	0.0222	0.0889
B	0.0667	0	0.0444	0.1111	0.0889	0.0667	0.0889	0.0667
C	0.0222	0.0444	0	0.1556	0.0444	0.0222	0.0444	0.1111
D	0.1333	0.1111	0.1556	0	0.2	0.1778	0.1111	0.0889
E	0.0667	0.0889	0.0444	0.2	0	0.0667	0.0889	0.1556
F	0.0444	0.0667	0.0222	0.1778	0.0667	0	0.0667	0.1333
G	0.0222	0.0889	0.0444	0.1111	0.0889	0.0667	0	0.1111
H	0.0889	0.0667	0.1111	0.0889	0.1556	0.1333	0.1111	0
I	0.0444	0.0667	0.0222	0.1333	0.0667	0.0444	0.0222	0.1333
J	0.1111	0.1778	0.1333	0.2	0.1778	0.1556	0.1333	0.1111
K	0.0889	0.0667	0.0667	0.1333	0.1111	0.0444	0.1111	0.0889
L	0.0444	0.0667	0.0222	0.1333	0.0667	0.0444	0.0222	0.1333
M	0.0444	0.0667	0.0222	0.1333	0.0667	0.0444	0.0222	0.1333
N	0.3333	0.3556	0.3556	0.2889	0.3556	0.3778	0.3556	0.2889
O	0.1111	0.1778	0.1333	0.1556	0.1778	0.1111	0.0889	0.1111
P	0.0444	0.0667	0.0222	0.1333	0.0667	0.0444	0.0222	0.1333
Q	0.7556	0.7333	0.7778	0.6667	0.7778	0.7556	0.7778	0.6667

Table 5 Results for non-membership values of overall expenditure of households in the Republic of Bulgaria (second part)

v-table	I	J	K	L	M	N	O	P	Q
A	0.0444	0.1111	0.0889	0.0444	0.0444	0.3333	0.1111	0.0444	0.7556
B	0.0667	0.1778	0.0667	0.0667	0.0667	0.3556	0.1778	0.0667	0.7333
C	0.0222	0.1333	0.0667	0.0222	0.0222	0.3556	0.1333	0.0222	0.7778
D	0.1333	0.2	0.1333	0.1333	0.1333	0.2889	0.1556	0.1333	0.6667
E	0.0667	0.1778	0.1111	0.0667	0.0667	0.3556	0.1778	0.0667	0.7778
F	0.0444	0.1556	0.0444	0.0444	0.0444	0.3778	0.1111	0.0444	0.7556
G	0.0222	0.1333	0.1111	0.0222	0.0222	0.3556	0.0889	0.0222	0.7778
H	0.1333	0.1111	0.0889	0.1333	0.1333	0.2889	0.1111	0.1333	0.6667
I	0	0.1556	0.0889	0	0	0.3778	0.1111	0	0.8
J	0.1556	0	0.1556	0.1556	0.1556	0.3111	0.1333	0.1556	0.6444
K	0.0889	0.1556	0	0.0889	0.0889	0.3333	0.1111	0.0889	0.7111
L	0	0.1556	0.0889	0	0	0.3778	0.1111	0	0.8
M	0	0.1556	0.0889	0	0	0.3778	0.1111	0	0.8
N	0.3778	0.3111	0.3333	0.3778	0.3778	0	0.3556	0.3778	0.4
O	0.1111	0.1333	0.1111	0.1111	0.1111	0.3556	0	0.1111	0.6889
P	0	0.1556	0.0889	0	0	0.3778	0.1111	0	0.8
Q	0.8	0.6444	0.7111	0.8	0.8	0.4	0.6889	0.8	0

3 Application of InterCriteria Analysis to Data on Overall Expenditure of Households in the Republic of Bulgaria

On the Fig. 2, is shown the analyzed data for the respective years. The Transpose function in the ICrAData software was used, which reorients the data for analysis from the perspective of different objects, in this case—years. The results for the degrees of membership are shown on the Table 5, the results for non-membership values are shown on the Table 6. The overall analysis for the respective years is shown on the Tables 7 and 8, from there it could be seen that all objects (years) corelate from strong positive to weak positive consonance (Table 9).

Analysis of Temporal Correlations Between Years

This analysis can aid in understanding economic trends and consumer behavior during the observed period:

- **Strong Positive Consonance [0.95; 1.00]** was noted for consecutive years such as 2014–2015, 2017–2018, and others up to 2022–2023. This indicates very strong relationships between expenditures in these years, which can be explained by stable economic conditions or unchanged consumer attitudes during these periods.
- **Positive Consonance [0.85; 0.95]** is observed between many years, indicating a strong but not as absolute correlation as the previous category. These years include broader intervals such as 2014–2023 and 2015–2023, highlighting strong but variable relationships over a longer period.

Table 6 Overall results for the expenditure of households in the Republic of Bulgaria

Type of correlations	Pairs of criteria	Number of pairs of criteria
Strong positive consonance [0,95; 1,00]	A-C, A-F, A-G, A-I, A-L, A-M, A-P, B-C, C-E, C-F, C-G, C-I, C-L, C-M, C-P, F-I, F-K, F-L, F-M, F-P, G-I, G-L, G-M, G-P, I-L, I-M, I-P, L-M, L-P, M-P	**30**
Positive consonance [0,85; 0,95)	A-B, A-D, A-E, A-H, A-J, A-K, A-O, B-D, B-E, B-F, B-G, B-H, B-I, B-K, B-L, B-M, B-P, C-H, C-J, C-K, C-O, D-G, D-H, D-I, D-K, D-L, D-M, D-P, E-F, E-G, E-I, E-K, E-L, E-M, E-P, F-G, F-H, F-O, G-H, G-J, G-K, G-O, H-I, H-J, H-K, H-L, H-M, H-O, H-P, I-K, I-O, J-O, K-L, K-M, K-O, K-P, L-O, M-O, O-P	**59**
Weak positive consonance [0,75; 0,85)	B-J, B-O, C-D, D-E, D-F, D-J, D-O, E-H, E-J, E-O, F-J, I-J, J-K, J-L, J-M, J-P	**16**
Weak dissonance [0,67; 0,75)	D-N, H-N	**2**
Dissonance [0,57; 0,67)	A-N, B-N, C-N, E-N, F-N, G-N, I-N, J-N, K-N, L-N, M-N, N-O, N-P, N-Q	14
Strong dissonance [0,43; 0,57)		–
Dissonance [0,33; 0,43)		–
Weak dissonance [0,25; 0,33)	D-Q, B-Q, H-Q, J-Q, K-Q, O-Q	6
Weak negative consonance [0,15;0,25)	A-Q, C-Q, E-Q, F-Q, G-Q, I-Q, L-Q, M-Q, P-Q	**9**
Negative consonance [0,05;0,15)		**–**
Strong negative consonance [0,00;0,05]		–

- **Weak Positive Consonance [0.75; 0.85]** for years like 2014–2020 and 2015–2021 shows a less pronounced but still significant connection. This may point to economic or social changes that affect consumer spending in the longer term.

Fig. 2 Analysis of expenditure data by year in ICrAData software

Table 7 Results for the membership values of overall expenditure of households in the Republic of Bulgaria, by year

μ-table	2014	2015	2016	2017	2018	2019	2020	2021	2022	2023
2014	1	0.9632	0.9338	0.875	0.875	0.8897	0.8309	0.8162	0.8529	0.8529
2015	0.9632	1	0.9265	0.8824	0.8824	0.8971	0.8382	0.8382	0.8603	0.875
2016	0.9338	0.9265	1	0.9118	0.9265	0.9191	0.8529	0.8382	0.875	0.875
2017	0.875	0.8824	0.9118	1	0.9706	0.9632	0.9118	0.9118	0.9191	0.9338
2018	0.875	0.8824	0.9265	0.9706	1	0.9632	0.8971	0.8824	0.9044	0.9191
2019	0.8897	0.8971	0.9191	0.9632	0.9632	1	0.9118	0.9118	0.9338	0.9485
2020	0.8309	0.8382	0.8529	0.9118	0.8971	0.9118	1	0.9559	0.9191	0.9191
2021	0.8162	0.8382	0.8382	0.9118	0.8824	0.9118	0.9559	1	0.9632	0.9485
2022	0.8529	0.8603	0.875	0.9191	0.9044	0.9338	0.9191	0.9632	1	0.9559
2023	0.8529	0.875	0.875	0.9338	0.9191	0.9485	0.9191	0.9485	0.9559	1

Table 8 Results for the non-membership values for the membership of overall expenditure of households in the Republic of Bulgaria, by year

ν-table	2014	2015	2016	2017	2018	2019	2020	2021	2022	2023
2014	0	0.0368	0.0662	0.125	0.125	0.1029	0.1691	0.1838	0.1471	0.1471
2015	0.0368	0	0.0735	0.1176	0.1176	0.0956	0.1618	0.1618	0.1397	0.125
2016	0.0662	0.0735	0	0.0882	0.0735	0.0735	0.1471	0.1618	0.125	0.125
2017	0.125	0.1176	0.0882	0	0.0294	0.0294	0.0882	0.0882	0.0809	0.0662
2018	0.125	0.1176	0.0735	0.0294	0	0.0294	0.1029	0.1176	0.0956	0.0809
2019	0.1029	0.0956	0.0735	0.0294	0.0294	0	0.0809	0.0809	0.0588	0.0441
2020	0.1691	0.1618	0.1471	0.0882	0.1029	0.0809	0	0.0441	0.0809	0.0809
2021	0.1838	0.1618	0.1618	0.0882	0.1176	0.0809	0.0441	0	0.0368	0.0515
2022	0.1471	0.1397	0.125	0.0809	0.0956	0.0588	0.0809	0.0368	0	0.0441
2023	0.1471	0.125	0.125	0.0662	0.0809	0.0441	0.0809	0.0515	0.0441	0

Table 9 The overall results for expenditures recorded over different years

Type of correlations	Pairs of criteria	Number of pairs of criteria
Strong positive consonance [0,95; 1,00]	2014–2015, 2017–2018, 2017–2019, 2018–2019, 2020–2021, 2021–2022, 2022–2023	**7**
Positive consonance [0,85; 0,95)	2014– 2016, 2014–2017,2014–2018, 2014–2019, 2014–2022, 2014–2023, 2015–2016, 2015–2017, 2015–2018, 2015–2019, 2015–2022, 2015–2023, 2016–2017, 2016–2018, 2016–2019, 2016 -2020, 2016 -2022, 2016 -2023 2017–2020, 2017–2021, 2017–2022, 2017–2023, 2018–2020, 2018–2021, 2018–2022, 2018–2023, 2019–2020, 2019–2021, 2019–2022, 2019–2023, 2020–2022,2020–2023, 2021–2023	**33**
Weak positive consonance [0,75; 0,85)	2014–2020, 2014–2021, 2015–2020, 2015–2021, 2016–2021	**5**

4 Conclusion

In this paper two analyses were applied over the data of overall household expenditure by expense groups in Republic of Bulgaria. The first analysis showed correlation in 7 different consonance groups and helps to better understand of the relationships between different categories of expenditures in this way it could support more effective planning of budgets and financial strategies for households. The analysis of correlations between the years 2014 to 2023 demonstrates various degrees of dependence in consumer spending or other economic indicators over the years. This type of analysis can be extremely useful for governmental bodies and economic analysts in planning budget strategies and policies aimed at stabilizing or stimulating consumption during certain periods.

References

1. Atanassov, K. (2014). *Index matrices towards an augmented matrix calculus. Studies in Computational Intelligence Series* (Vol. 573). Springer
2. Atanassov, K. (2012). *On intuitionistic fuzzy sets theory*. Springer.
3. Atanassov, K., Mavrov, D., & Atanassova, V. (2014). InterCriteria decision making: A new approach for multicriteria decision making, based on index matrices and intuitionistic fuzzy sets. *Issues in Intuitionistic Fuzzy Sets and Generalized Nets, 11*, 1–8.
4. Atanassov, K., Szmidt, E., & Kacprzyk, J. (2013). On intuitionistic fuzzy pairs. *Notes on Intuitionistic Fuzzy Sets, 19*(3), 1–13.
5. Atanassova, V., Doukovska, L., Atanassov, K., Mavrov, D.: InterCriteria decision making approach to EU member states competitiveness analysis. In *Proc of the international symposium on business modeling and software design—BMSD'14, 24–26 June 2014, Luxembourg, Grand Duchy of Luxembourg* (pp. 289–294)
6. Bureva, V., & Krawczak, M. (2024). InterCriteria Analysis as an intelligent tool for decision making: Investigation of Polish University Rankings. *Notes on Intuitionistic Fuzzy Sets, 30*(2), 180–189.
7. Chorukova, E., Marinov, P., & Umlenski, I. (2021). Survey on theory and applications of intercriteria analysis approach. In *Atanassov, K.T. (ed.). Research in Computer Science in the Bulgarian Academy of Sciences. Studies in Computational Intelligence* (Vol. 934). Springer. https://doi.org/10.1007/978-3-030-72284-520
8. Ikonomov, N., Vassilev, P., Roeva, O. (2018). ICrAData—software for intercriteria analysis. *Int J Bioautomation, 22*(1), 1–10.
9. Ikonomov, N., Vassilev, P., & Roeva, O. (2018). ICrAData—software for intercriteria analysis. *International Journal Bioautomation, 22*(1), 1–10.
10. Krawczak, M., Bureva, V., Sotirova, E., & Szmidt, E. (2016). Application of the InterCriteria decision making method to universities ranking. In: *Novel developments in uncertainty representation and processing, advances in intelligent systems and computing* (Vol. 401, pp. 365–372). Springer
11. National Statistical Institute of the Republic of Bulgaria: https://nsi.bg/bg/content/3239/%D0%B3%D0%BE%D0%B4%D0%B8%D1%88%D0%BD%D0%B8-%D0%B4%D0%B0%D0%BD%D0%BD%D0%B8

Neural Networks

An Application of InterCriteria Analysis with Self Organizing Map Neural Network

Todor Petkov

Abstract In this paper an application of InterCriteria Analysis (ICA) with self organizing map neural network is presented. ICA is based on the apparatus of the index matrices and the intuitionistic fuzzy sets. The data for analysis represents proportion of population for all of the European Union countries plus other four from Europe. The analysis is applied over the countries and then membership and non-membership values are used to train two Self Organizing Map neural networks.

Keywords Artificial neural network · Self organizing map · InterCriteria analysis

1 Introduction

InterCriteria Analysis gives possibility to compare some criteria or estimated by their objects. By this reason it is called an intercriteria decision making method [3–5, 12]. Let an Index Matrix (IM) [1, 2] whose index sets consist of the names of the criteria (for rows) and objects (for columns) be given. An IM with index sets consisting of the names of the criteria (for rows and for columns) with elements IFPs corresponding to the "agreement" and "disagreement" of the respective criteria will be obtained [6, 11, 12].

The set of all objects being evaluated is denoted by O, and the set of values assigned by a given criteria C to the objects by $C(O)$:

$$O \stackrel{\text{def}}{=} \{O_1, O_2, ..., O_n\}, C(O) \stackrel{\text{def}}{=} \{C(O_1), C(O_2), ..., C(O_n)\}.$$

T. Petkov (✉)
Department of Computer Systems and Technologies, University prof. Asen Zlatarov, Burgas, Bulgaria
e-mail: todor_petkov@btu.bg

O. Castillo and P. Melin (eds.), *Intelligent System Design Based on Soft Computing Models*, Studies in Computational Intelligence 1223,
https://doi.org/10.1007/978-3-031-97309-3_7

According to [3] in order to construct Index Matrix (IM) with criterias C:

$$\begin{array}{c|ccc} & C_1 & \dots & C_m \\ \hline C_1 & \langle \mu C_1, C_1, \nu C_1, C_1 \rangle & \dots & \langle \mu C_1, C_m, \nu C_1, C_m \rangle \\ \vdots & \vdots & & \vdots \\ C_m & \langle \mu C_m, C_1, \nu C_m, C_1 \rangle & \dots & \langle \mu C_m, C_m, \nu C_m, C_m \rangle \end{array}$$

that determine the degrees of correspondence between all of the criteria:

- μ is the degree of membership
- ν is the degree of non-membership.

The criteria C_k and C_l are:

- in positive consonance if $\mu C_k, C_l > \alpha$ and $\nu\ C_k, C_l < \beta$
- in negative consonance if $\mu C_k, C_l < \beta$ and $\mu\ C_k, C_l > \alpha$
- dissonance otherwise.

The main basic building block of an ANN [7, 8] is the Artificial Neuron shown in Fig. 1. The artificial neuron tries to represent the work of the biological neuron. Each artificial neuron has an input(s) and an output; the inputs are the incoming signals from the other neurons and the output is the result from the neuron to the next neuron(s). The inputs to the neuron in Fig. 1 are denoted as p where p is a vector of the incoming signals:

$$p = [p_1, p_2, \dots, p_k]$$

where k is the number of incoming signals.

The scalar input p and scalar weights w are multiplied to form the input to the summer block the summer output value is goes into the transfer function f which produce the scalar neuron output a.

$$a = f\left(\sum_{i=0}^{k} p_i w_i\right)$$

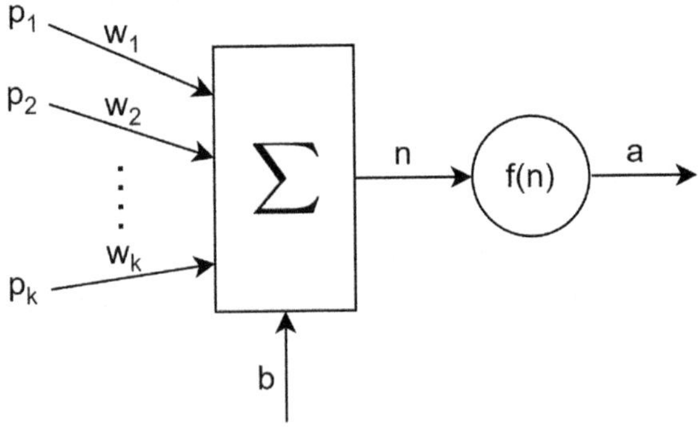

Fig. 1 The model of artificial neuron

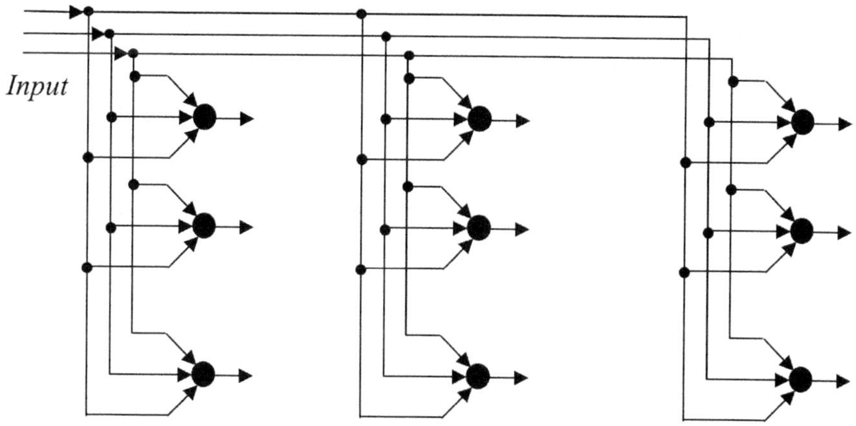

Fig. 2 The general structure of self organizing map neural network

The two main types of learning of artifical neural networks can be categorized as supervised and unsupervised [9]. In supervised learning for each input vector has to be provided target and the network tries to minimize the error between the actual output and the real one that was provided.

On the other hand in unsupervised learning the input vectors do not have target ones and the network tries group(clustering) the input data in based on the similarity of the features of the input data. The Self Organizing Map (SOM) Fig. 2 is unsupervised competitive neural network where each neuron competes for the input data and the winner neuron will change its weights along with its neighbouring neurons of a certain distance.

2 Task Essentials

InterCriteria Analysis has been applied to data that was taken from [13]. The data represents the proportion of population (by percentages) of the all European Union (EU) countries plus Iceland, Liechtenstein, Norway and Switzerland. The people are divided into ten groups by age as follows:

- less than 15 years
- from 15 to 29 years
- from 30 to 34 years
- from 35 to 39 years
- from 40 to 44 years
- from 45 to 49 years
- from 50 to 54 years
- from 55 to 59 years

Table 1 Proportion population (by %) in Europe

	Less than 15 years	15 to 29 years	30 to 34 years	35 to 39 years	40 to 44 years	45 to 49 years	50 to 54 years	55 to 59 years	60 to 64 years	65 years or over
Belgium	14.6	18.2	9.0	8.6	8.6	8.3	8.0	6.9	5.3	12.6
Bulgaria	3.1	13.1	8.5	8.6	9.5	10.3	9.5	9.2	7.8	20.4
Czechia	10.1	13.3	13.4	15.7	14.1	10.4	7.0	5.1	3.8	7.1
Denmark	12.8	27.5	14.4	12.2	9.5	7.0	5.5	3.9	2.6	4.5
Germany	12.2	19.8	10.5	9.7	9.5	9.5	8.3	6.1	4.6	9.9
Estonia	7.6	28.3	11.1	10.0	8.6	7.9	6.9	5.7	4.3	9.5
Ireland	12.1	17.8	11.5	17.3	16.6	10.1	5.3	3.5	2.9	2.9
Greece	8.0	13.4	6.9	8.8	11.2	12.2	11.2	8.8	7.3	12.3
Spain	12.0	15.4	9.1	10.7	11.5	10.1	8.8	6.3	5.1	11.0
France	15.1	13.7	6.3	6.9	7.4	7.5	8.3	8.1	6.4	20.3
Croatia	5.8	13.4	5.3	6.8	8.5	8.1	7.7	6.8	7.4	30.3
Italy	16.4	13.6	9.1	10.9	12.1	10.9	10.5	6.5	5.0	4.8
Cypris	12.0	19.5	14.5	13.1	11.0	8.5	6.0	4.8	4.2	6.5
Latvia	7.0	12.6	10.1	10.5	9.6	9.9	10.4	8.6	6.6	14.8
Lithuania	7.3	18.4	14.6	15.7	13.2	9.6	7.0	4.5	3.3	6.4
Luxembourg	15.8	15.3	8.5	9.1	9.7	9.6	8.8	7.5	5.3	10.5
Hungary	6.7	23.4	10.3	8.2	7.4	7.1	7.6	6.7	6.9	15.6
Malta	10.1	24.8	18.3	13.3	9.5	6.6	5.0	3.5	2.4	6.3
Netherlands	11.8	29.0	13.3	11.6	9.5	7.5	5.9	4.2	2.7	4.5
Austria	15.5	20.2	11.0	10.7	10.0	8.5	7.5	5.9	4.0	6.8
Poland	6.2	14.0	13.4	12.8	10.7	8.4	7.2	5.3	5.3	16.5
Portugal	6.9	20.8	10.9	9.8	9.1	7.4	6.7	5.9	6.1	16.3
Romania	14.8	15.0	7.4	7.1	7.5	7.8	8.1	5.9	6.4	20.1
Slovenia	6.5	12.2	8.4	10.7	11.2	10.8	9.9	9.4	6.8	14.2
Slovakia	4.4	12.3	12.2	13.8	12.6	10.9	9.1	7.0	5.0	12.6
Finland	14.4	15.9	11.6	12.6	11.0	9.2	8.0	6.2	4.7	6.3
Sweden	12.6	18.2	12.2	11.2	9.6	7.8	6.4	5.1	4.1	12.9
Iceland	11.0	29.0	16.3	12.8	9.6	7.5	5.6	3.8	2.5	1.9
Liechtenstein	11.0	12.7	5.7	7.3	8.8	10.0	11.5	11.2	7.6	14.1
Norway	15.3	15.7	12.2	13.8	12.5	9.9	7.6	5.4	3.5	4.2
Switzerland	14.3	15.9	10.2	10.9	10.4	9.1	8.4	7.1	4.3	9.5

- from 60 to 64 years
- 65 years or over.

The period of the data collection is from 2021 till 2023 and covers the 31 countries that were mentioned. In the Table 1 is shown the data that is represented by percentages for the population for that countries.

Table 2 Abbreviation for the countries in Europe

Belgium	BE	Hungary	HU
Bulgaria	BG	Malta	MT
Czechia	CZ	Netherlands	NL
Denmark	DK	Austria	AT
Germany	DE	Poland	PL
Estonia	EE	Portugal	PT
Ireland	IE	Romania	RO
Greece	GR	Slovenia	SI
Spain	ES	Slovakia	SK
France	FR	Finland	FI
Croatia	HR	Sweden	SE
Italy	IT	Iceland	IS
Cyprus	CY	Liechtenstein	LI
Latvia	LV	Norway	NO
Lithuania	LT	Switzerland	CH
Luxembourg	LU		

The InterCriteria Analysis was applied over the data from Table 1 using the ICrA-Data software. The initial analysis aimed to establish the relationships between all of the EU Countries. The data were entered into ICrAData with default options (Standart ICrA method and μ–biased ICrA Variant) for analysis. In order to simplify the abbreviation of the countries their domain names are used shown on the Table 2.

The μ–analysis analysis was applied over all of the countries. The output matrix from the analysis is large and the μ–analysis is applied as an appendix here and also is divided into three parts for first ten countries along with other 31 5, then for the second part the 31 countries and the next ten countries (Croatia, Italy, Cyprus, Latvia, Lithuania, Luxembourg, Hungary, Malta, Netherlands and Austria) are shown on the Table 6 and the third last part of the analysis between all 31 countries and the next eleven countries (Poland, Portugal, Romania, Slovenia, Slovakia, Finland, Sweden, Iceland, Liechtenstein, Norway and Switzerland) are shown on the Table 7. The μ and ν analysis over all of the available data(countries) is shown on the Fig. 4.

The fuzzy pairs from the ICA analysis are grouped by using the self organizing map neural network this is used in order to simplify the analysis and the countries that correlates can be a part of a cluster and those cluster is their representative sample. The most important at the beginning is the number of the clusters here the SOM is divided into 5 clusters an the results, the input vector consists of 31 values that are in the range between 0 and 1 the results are shown on the Fig. 5a and b. It can be seen from the results from the Fig. 5 the training set is clustered among all of the clusters and on the Table 3 is shown the μ statistic result.

The ν analysis for all 31 countries is made and also is applied as an appendix here at the end of the paper the ν–values from the ICA analysis are shown on the

Table 3 The μ data statistic for the countries

Cluster #	Countries across clusters
Cluster 1	BG; GR; FR; HR; LI
Cluster 2	LV; HU; RO; SI; SK
Cluster 3	BE; CZ; ES; IT; LU; PL; PT
Cluster 4	DK; DE; EE; IE; CY; LT; NL; AT; FI; SE; IS; CH
Cluster 5	MT; NO

Table 4 The ν data statistic for the countries

Cluster #	Countries across clusters
Cluster 1	BG; GR; FR; HR; LI
Cluster 2	LV; RO; SI
Cluster 3	LU; HU; PL; PT; SK
Cluster 4	BE; CZ; DE; EE; ES; SE
Cluster 5	DK; IE; IT; CY; LT; MT; NL; AT; FI; IS; NO; CH

Tables 8, 9 and 10. The ν fuzzy pairs for is also used as an inputs for the SOM neural network. The SOM is also divided into 5 clusters and each of them is the representative sample of the group of the countries that correlates by using their ν–values. On the Fig. 3a and b are shown the structure and the sample hits of the training set for the non-membership values from the ICA analysis. In the ν analysis the statistics for the cluster is also shown on the Table 4.

3 Conclusion

An application of the InterCriteria Analysis with Self Organizing Map neural network was presented over the data that represents proportion of the population in Europe. This correlation analysis helps to better understanding the relationships between different countries of their proportion population. The proportion of population for each of the countries differs and the analysis is a tool not only for identifying the relationships between them but also could be use as a tool that helps to find the solutions for similar population problems. The use of SOM NN is useful here in order to identify groups of the similar countries and in this way to reduce the use of the big dataset for representation. Each of the group (cluster) is the representative sample of the countries that belongs to it and different solutions for population problems for each of the groups could be applied.

4 Appendix

See Tables 5, 6, 7, 8, 9 and 10.

Table 5 μ analysis of proportion population (by %) in Europe (1–10)

	BE	BG	CZ	DK	DE	EE	IE	GR	ES	FR
BE	–	0.56	0.69	0.84	0.93	0.82	0.73	0.60	0.89	0.60
BG	0.56	–	0.53	0.47	0.53	0.62	0.49	0.89	0.62	0.71
CZ	0.69	0.53	–	0.80	0.71	0.82	0.82	0.53	0.73	0.40
DK	0.84	0.47	0.80	–	0.87	0.84	0.84	0.47	0.80	0.47
DE	0.93	0.53	0.71	0.87	–	0.84	0.76	0.58	0.84	0.58
EE	0.82	0.62	0.82	0.84	0.84	–	0.78	0.62	0.73	0.49
IE	0.73	0.49	0.82	0.84	0.76	0.78	–	0.53	0.78	0.44
GR	0.60	0.89	0.53	0.47	0.58	0.62	0.53	–	0.67	0.73
ES	0.89	0.62	0.73	0.80	0.84	0.73	0.78	0.67	–	0.67
FR	0.60	0.71	0.40	0.47	0.58	0.49	0.44	0.73	0.67	–
HR	0.56	0.84	0.49	0.42	0.51	0.58	0.49	0.87	0.62	0.67
IT	0.69	0.49	0.67	0.78	0.69	0.62	0.82	0.53	0.80	0.58
CY	0.84	0.51	0.84	0.96	0.87	0.89	0.84	0.51	0.80	0.47
LV	0.69	0.78	0.64	0.62	0.69	0.78	0.60	0.73	0.64	0.62
LT	0.76	0.56	0.89	0.91	0.78	0.89	0.89	0.56	0.76	0.42
LU	0.84	0.67	0.64	0.71	0.80	0.69	0.69	0.71	0.91	0.76
HU	0.73	0.67	0.64	0.67	0.71	0.82	0.60	0.62	0.64	0.49
MT	0.84	0.51	0.84	0.96	0.87	0.89	0.84	0.51	0.80	0.47
NL	0.84	0.47	0.80	1.00	0.87	0.84	0.84	0.47	0.80	0.47
AT	0.87	0.44	0.78	0.98	0.89	0.82	0.87	0.49	0.82	0.49
PL	0.80	0.69	0.71	0.73	0.78	0.89	0.67	0.64	0.71	0.51
PT	0.84	0.64	0.76	0.78	0.82	0.93	0.71	0.64	0.76	0.51
RO	0.73	0.71	0.49	0.60	0.71	0.62	0.53	0.71	0.76	0.82
SI	0.64	0.91	0.62	0.56	0.60	0.71	0.58	0.82	0.71	0.64
SK	0.64	0.71	0.80	0.64	0.62	0.78	0.67	0.62	0.67	0.44
FI	0.84	0.47	0.80	0.96	0.87	0.80	0.89	0.51	0.84	0.51
SE	0.96	0.58	0.69	0.84	0.93	0.87	0.73	0.62	0.87	0.62
IS	0.78	0.44	0.78	0.93	0.80	0.82	0.89	0.44	0.73	0.40
LI	0.56	0.76	0.36	0.42	0.53	0.44	0.40	0.78	0.62	0.91
NO	0.80	0.47	0.80	0.91	0.82	0.76	0.93	0.51	0.84	0.51
CH	0.87	0.53	0.82	0.89	0.89	0.82	0.87	0.58	0.91	0.58

Table 6 μ analysis of proportion population (by %) in Europe (11–20)

	HR	IT	CY	LV	LT	LU	HU	MT	NL	AT
BE	0.56	0.69	0.84	0.69	0.76	0.84	0.73	0.84	0.84	0.87
BG	0.84	0.49	0.51	0.78	0.56	0.67	0.67	0.51	0.47	0.44
CZ	0.49	0.67	0.84	0.64	0.89	0.64	0.64	0.84	0.80	0.78
DK	0.42	0.78	0.96	0.62	0.91	0.71	0.67	0.965	1.00	0.98
DE	0.51	0.69	0.87	0.69	0.78	0.80	0.71	0.87	0.87	0.89
EE	0.58	0.62	0.89	0.78	0.89	0.69	0.82	0.89	0.84	0.82
IE	0.49	0.82	0.84	0.60	0.89	0.69	0.60	0.84	0.84	0.87
GR	0.87	0.53	0.51	0.73	0.56	0.71	0.62	0.51	0.47	0.49
ES	0.62	0.80	0.80	0.64	0.76	0.91	0.64	0.80	0.80	0.82
FR	0.67	0.58	0.47	0.62	0.42	0.76	0.49	0.47	0.47	0.49
HR	–	0.49	0.47	0.67	0.51	0.67	0.64	0.47	0.42	0.44
IT	0.49	–	0.73	0.49	0.73	0.82	0.44	0.73	0.78	0.80
CY	0.47	0.73	–	0.67	0.91	0.71	0.71	1.00	0.96	0.93
LV	0.67	0.49	0.67	–	0.71	0.64	0.84	0.67	0.62	0.60
LT	0.51	0.73	0.91	0.71	–	0.67	0.71	0.91	0.91	0.89
LU	0.67	0.82	0.71	0.64	0.67	–	0.60	0.71	0.71	0.73
HU	0.64	0.44	0.71	0.84	0.71	0.60	–	0.71	0.67	0.64
MT	0.47	0.73	1.00	0.67	0.91	0.71	0.71	–	0.96	0.93
NL	0.42	0.78	0.96	0.62	0.91	0.71	0.67	0.96	–	0.98
AT	0.44	0.80	0.93	0.60	0.89	0.73	0.64	0.93	0.98	–
PL	0.67	0.51	0.78	0.84	0.78	0.67	0.87	0.78	0.73	0.71
PT	0.64	0.56	0.82	0.80	0.82	0.71	0.89	0.82	0.78	0.76
RO	0.69	0.62	0.60	0.71	0.51	0.80	0.67	0.60	0.60	0.62
SI	0.87	0.58	0.60	0.80	0.64	0.76	0.73	0.60	0.56	0.53
SK	0.67	0.51	0.69	0.78	0.73	0.60	0.76	0.69	0.64	0.62
FI	0.47	0.82	0.91	0.62	0.91	0.76	0.62	0.91	0.96	0.98
SE	0.58	0.67	0.84	0.73	0.76	0.82	0.78	0.84	0.84	0.87
IS	0.40	0.80	0.93	0.60	0.89	0.64	0.64	0.93	0.93	0.91
LI	0.76	0.49	0.42	0.67	0.38	0.67	0.51	0.42	0.42	0.44
NO	0.47	0.87	0.87	0.58	0.87	0.76	0.58	0.87	0.91	0.93
CH	0.53	0.80	0.89	0.64	0.84	0.82	0.64	0.89	0.89	0.91

Table 7 μ analysis of proportion population (by %) in Europe (21–31)

	PL	PT	RO	SI	SK	FI	SE	IS	LI	NO	CH
BE	0.80	0.84	0.73	0.64	0.64	0.84	0.96	0.78	0.56	0.80	0.87
BG	0.69	0.64	0.71	0.91	0.71	0.47	0.58	0.44	0.76	0.47	0.53
CZ	0.71	0.76	0.49	0.62	0.80	0.80	0.69	0.78	0.36	0.80	0.82
DK	0.73	0.78	0.60	0.56	0.64	0.96	0.84	0.93	0.42	0.91	0.89
DE	0.78	0.82	0.71	0.60	0.62	0.87	0.93	0.80	0.53	0.82	0.89
EE	0.89	0.93	0.62	0.71	0.78	0.80	0.87	0.82	0.44	0.76	0.82
IE	0.67	0.71	0.53	0.58	0.67	0.89	0.73	0.89	0.40	0.93	0.87
GR	0.64	0.64	0.71	0.82	0.62	0.51	0.62	0.44	0.78	0.51	0.58
ES	0.71	0.76	0.76	0.71	0.67	0.84	0.87	0.73	0.62	0.84	0.91
FR	0.51	0.51	0.82	0.64	0.44	0.51	0.62	0.40	0.91	0.51	0.58
HR	0.67	0.64	0.69	0.87	0.67	0.47	0.58	0.40	0.76	0.47	0.53
IT	0.51	0.56	0.62	0.58	0.51	0.82	0.67	0.80	0.49	0.87	0.80
CY	0.78	0.82	0.60	0.60	0.69	0.91	0.84	0.93	0.42	0.87	0.89
LV	0.84	0.80	0.71	0.80	0.78	0.62	0.73	0.60	0.67	0.58	0.64
LT	0.78	0.82	0.51	0.64	0.73	0.91	0.76	0.89	0.38	0.87	0.84
LU	0.67	0.71	0.80	0.76	0.60	0.76	0.82	0.64	0.67	0.76	0.82
HU	0.87	0.89	0.67	0.73	0.76	0.62	0.78	0.64	0.51	0.58	0.64
MT	0.78	0.82	0.60	0.60	0.69	0.91	0.84	0.93	0.42	0.87	0.89
NL	0.73	0.78	0.60	0.56	0.64	0.96	0.84	0.93	0.42	0.91	0.89
AT	0.71	0.76	0.62	0.53	0.62	0.98	0.87	0.91	0.44	0.93	0.91
PL	–	0.93	0.67	0.78	0.80	0.69	0.84	0.71	0.51	0.64	0.71
PT	0.93	–	0.69	0.73	0.76	0.73	0.89	0.76	0.47	0.69	0.76
RO	0.67	0.69	–	0.69	0.53	0.60	0.76	0.53	0.78	0.60	0.67
SI	0.78	0.73	0.69	–	0.80	0.56	0.67	0.53	0.69	0.56	0.62
SK	0.80	0.76	0.53	0.80	–	0.64	0.69	0.62	0.49	0.64	0.71
FI	0.69	0.73	0.60	0.56	0.64	–	0.84	0.89	0.47	0.96	0.93
SE	0.84	0.89	0.76	0.67	0.69	0.84	–	0.78	0.58	0.80	0.87
IS	0.71	0.76	0.53	0.53	0.62	0.89	0.78	–	0.36	0.89	0.82
LI	0.51	0.47	0.78	0.69	0.49	0.47	0.58	0.36	–	0.47	0.53
NO	0.64	0.69	0.60	0.56	0.64	0.96	0.80	0.89	0.47	–	0.93
CH	0.71	0.76	0.67	0.62	0.71	0.93	0.87	0.82	0.53	0.93	–

Table 8 ν analysis of proportion population (by %) in Europe (1–10)

	BE	BG	CZ	DK	DE	EE	IE	GR	ES	FR
BE	–	0.40	0.29	0.13	0.02	0.16	0.22	0.33	0.09	0.38
BG	0.40	–	0.44	0.51	0.42	0.36	0.47	0.09	0.36	0.27
CZ	0.29	0.44	–	0.20	0.27	0.18	0.16	0.42	0.27	0.60
DK	0.13	0.51	0.20	–	0.11	0.16	0.13	0.49	0.20	0.53
DE	0.02	0.42	0.27	0.11	–	0.13	0.20	0.36	0.13	0.40
EE	0.16	0.36	0.18	0.16	0.13	–	0.20	0.33	0.27	0.51
IE	0.22	0.47	0.16	0.13	0.20	0.20	–	0.40	0.20	0.53
GR	0.33	0.09	0.42	0.49	0.36	0.33	0.40	–	0.29	0.22
ES	0.09	0.36	0.27	0.20	0.13	0.27	0.20	0.29	–	0.33
FR	0.38	0.27	0.60	0.53	0.40	0.51	0.53	0.22	0.33	–
HR	0.40	0.11	0.49	0.56	0.44	0.40	0.47	0.11	0.36	0.31
IT	0.27	0.47	0.31	0.20	0.27	0.36	0.13	0.40	0.18	0.40
CY	0.13	0.47	0.16	0.04	0.11	0.11	0.13	0.44	0.20	0.53
LV	0.29	0.20	0.36	0.38	0.29	0.22	0.38	0.22	0.36	0.38
LT	0.22	0.42	0.11	0.09	0.20	0.11	0.09	0.40	0.24	0.58
LU	0.13	0.31	0.36	0.29	0.18	0.31	0.29	0.24	0.09	0.24
HU	0.22	0.29	0.33	0.31	0.24	0.16	0.36	0.31	0.33	0.49
MT	0.13	0.47	0.16	0.04	0.11	0.11	0.13	0.44	0.20	0.53
NL	0.13	0.51	0.20	0.00	0.11	0.16	0.13	0.49	0.20	0.53
AT	0.11	0.53	0.22	0.02	0.09	0.18	0.11	0.47	0.18	0.51
PL	0.16	0.27	0.27	0.24	0.18	0.09	0.29	0.29	0.27	0.47
PT	0.13	0.33	0.24	0.22	0.16	0.07	0.27	0.31	0.24	0.49
RO	0.24	0.27	0.51	0.40	0.27	0.38	0.44	0.24	0.24	0.18
SI	0.33	0.07	0.38	0.44	0.38	0.29	0.40	0.13	0.29	0.36
SK	0.31	0.24	0.18	0.33	0.33	0.20	0.29	0.31	0.31	0.53
FI	0.13	0.51	0.20	0.04	0.11	0.20	0.09	0.44	0.16	0.49
SE	0.02	0.40	0.31	0.16	0.04	0.13	0.24	0.33	0.13	0.38
IS	0.20	0.53	0.22	0.07	0.18	0.18	0.09	0.51	0.27	0.60
LI	0.42	0.22	0.64	0.58	0.44	0.56	0.58	0.18	0.38	0.09
NO	0.18	0.51	0.20	0.09	0.16	0.24	0.04	0.44	0.16	0.49
CH	0.11	0.44	0.18	0.11	0.09	0.18	0.11	0.38	0.09	0.42

Table 9 ν analysis of proportion population (by %) in Europe (11–20)

	HR	IT	CY	LV	LT	LU	HU	MT	NL	AT
BE	0.40	0.27	0.13	0.29	0.22	0.13	0.22	0.13	0.13	0.11
BG	0.11	0.47	0.47	0.20	0.42	0.31	0.29	0.47	0.51	0.53
CZ	0.49	0.31	0.16	0.36	0.11	0.36	0.33	0.16	0.20	0.22
DK	0.56	0.20	0.04	0.38	0.09	0.29	0.31	0.04	0.00	0.02
DE	0.44	0.27	0.11	0.29	0.20	0.18	0.24	0.11	0.11	0.09
EE	0.40	0.36	0.11	0.22	0.11	0.31	0.16	0.11	0.16	0.18
IE	0.47	0.13	0.13	0.38	0.09	0.29	0.36	0.13	0.13	0.11
GR	0.11	0.40	0.44	0.22	0.40	0.24	0.31	0.44	0.49	0.47
ES	0.36	0.18	0.20	0.36	0.24	0.09	0.33	0.20	0.20	0.18
FR	0.31	0.40	0.53	0.38	0.58	0.24	0.49	0.53	0.53	0.51
HR	–	0.47	0.51	0.31	0.47	0.31	0.31	0.51	0.56	0.53
IT	0.47	–	0.24	0.49	0.24	0.16	0.51	0.24	0.20	0.18
CY	0.51	0.24	–	0.33	0.09	0.29	0.27	0.00	0.04	0.07
LV	0.31	0.49	0.33	–	0.29	0.36	0.13	0.33	0.38	0.40
LT	0.47	0.24	0.09	0.29	–	0.33	0.27	0.09	0.09	0.11
LU	0.31	0.16	0.29	0.36	0.33	–	0.38	0.29	0.29	0.27
HU	0.31	0.51	0.27	0.13	0.27	0.38	–	0.27	0.31	0.33
MT	0.51	0.24	0.00	0.33	0.09	0.29	0.27	–	0.04	0.07
NL	0.56	0.20	0.04	0.38	0.09	0.29	0.31	0.04	–	0.02
AT	0.53	0.18	0.07	0.40	0.11	0.27	0.33	0.07	0.02	–
PL	0.29	0.44	0.20	0.13	0.20	0.31	0.09	0.20	0.24	0.27
PT	0.33	0.42	0.18	0.20	0.18	0.29	0.09	0.18	0.22	0.24
RO	0.29	0.36	0.40	0.29	0.49	0.20	0.31	0.40	0.40	0.38
SI	0.11	0.40	0.40	0.20	0.36	0.24	0.24	0.40	0.44	0.47
SK	0.29	0.44	0.29	0.20	0.24	0.38	0.20	0.29	0.33	0.36
FI	0.51	0.16	0.09	0.38	0.09	0.24	0.36	0.09	0.04	0.02
SE	0.40	0.31	0.16	0.27	0.24	0.18	0.20	0.16	0.16	0.13
IS	0.58	0.18	0.07	0.40	0.11	0.36	0.33	0.07	0.07	0.09
LI	0.22	0.49	0.58	0.33	0.62	0.33	0.47	0.58	0.58	0.56
NO	0.51	0.11	0.13	0.42	0.13	0.24	0.40	0.13	0.09	0.07
CH	0.44	0.18	0.11	0.36	0.16	0.18	0.33	0.11	0.11	0.09

Table 10 ν analysis of proportion population (by %) in Europe (21–31)

	PL	PT	RO	SI	SK	FI	SE	IS	LI	NO	CH
BE	0.16	0.13	0.24	0.33	0.31	0.13	0.02	0.20	0.42	0.18	0.11
BG	0.27	0.33	0.27	0.07	0.24	0.51	0.40	0.53	0.22	0.51	0.44
CZ	0.27	0.24	0.51	0.38	0.18	0.20	0.31	0.22	0.64	0.20	0.18
DK	0.24	0.22	0.40	0.44	0.33	0.04	0.16	0.07	0.58	0.09	0.11
DE	0.18	0.16	0.27	0.38	0.33	0.11	0.04	0.18	0.44	0.16	0.09
EE	0.09	0.07	0.38	0.29	0.20	0.20	0.13	0.18	0.56	0.24	0.18
IE	0.29	0.27	0.44	0.40	0.29	0.09	0.24	0.09	0.58	0.04	0.11
GR	0.29	0.31	0.24	0.13	0.31	0.44	0.33	0.51	0.18	0.44	0.38
ES	0.27	0.24	0.24	0.29	0.31	0.16	0.13	0.27	0.38	0.16	0.09
FR	0.47	0.49	0.18	0.36	0.53	0.49	0.38	0.60	0.09	0.49	0.42
HR	0.29	0.33	0.29	0.11	0.29	0.51	0.40	0.58	0.22	0.51	0.44
IT	0.44	0.42	0.36	0.40	0.44	0.16	0.31	0.18	0.49	0.11	0.18
CY	0.20	0.18	0.40	0.40	0.29	0.09	0.16	0.07	0.58	0.13	0.11
LV	0.13	0.20	0.29	0.20	0.20	0.38	0.27	0.40	0.33	0.42	0.36
LT	0.20	0.18	0.49	0.36	0.24	0.09	0.24	0.11	0.62	0.13	0.16
LU	0.31	0.29	0.20	0.24	0.38	0.24	0.18	0.36	0.33	0.24	0.18
HU	0.09	0.09	0.31	0.24	0.20	0.36	0.20	0.33	0.47	0.40	0.33
MT	0.20	0.18	0.40	0.40	0.29	0.09	0.16	0.07	0.58	0.13	0.11
NL	0.24	0.22	0.40	0.44	0.33	0.04	0.16	0.07	0.58	0.09	0.11
AT	0.27	0.24	0.38	0.47	0.36	0.02	0.13	0.09	0.56	0.07	0.09
PL	–	0.04	0.31	0.20	0.16	0.29	0.13	0.27	0.47	0.33	0.27
PT	0.04	–	0.31	0.27	0.22	0.27	0.11	0.24	0.53	0.31	0.24
RO	0.31	0.31	–	0.31	0.44	0.40	0.24	0.47	0.22	0.40	0.33
SI	0.20	0.27	0.31	–	0.18	0.44	0.33	0.47	0.31	0.44	0.38
SK	0.16	0.22	0.44	0.18	–	0.33	0.29	0.36	0.49	0.33	0.27
FI	0.29	0.27	0.40	0.44	0.33	–	0.16	0.11	0.53	0.04	0.07
SE	0.13	0.11	0.24	0.33	0.29	0.16	–	0.22	0.42	0.20	0.13
IS	0.27	0.24	0.47	0.47	0.36	0.11	0.22	–	0.64	0.11	0.18
LI	0.47	0.53	0.22	0.31	0.49	0.53	0.42	0.64	–	0.53	0.47
NO	0.33	0.31	0.40	0.44	0.33	0.04	0.20	0.11	0.53	–	0.07
CH	0.27	0.24	0.33	0.38	0.27	0.07	0.13	0.18	0.47	0.07	–

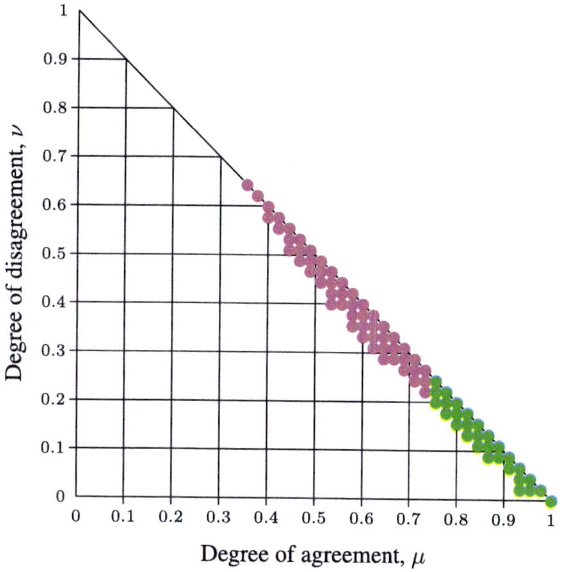

Fig. 3 μ and ν analysis over the countries

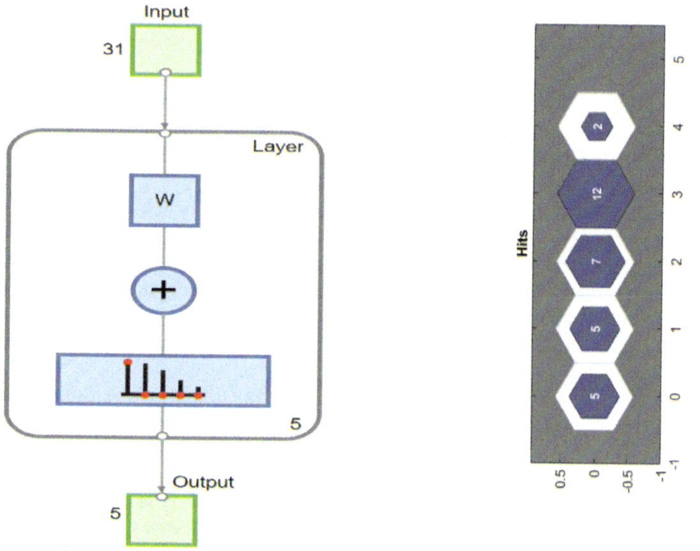

(a) Structure of SOM Neural Network

(b) Sample hits of the training set along the clusters

Fig. 4 μ–analysis SOM structure and results

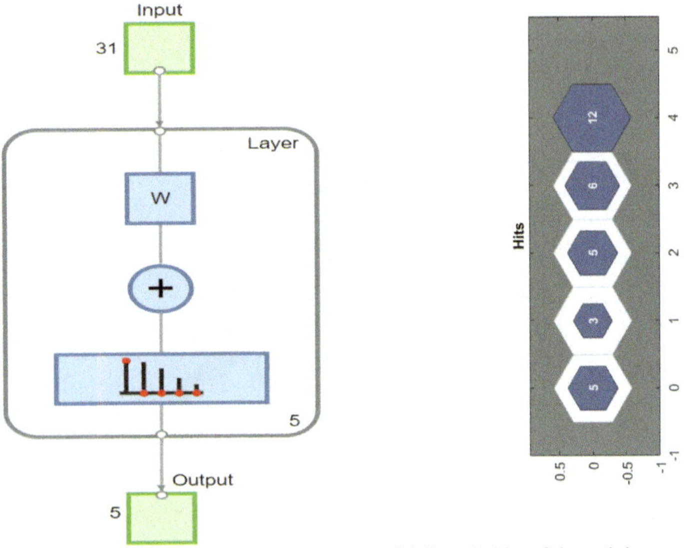

(a) Structure of SOM Neural Network

(b) Sample hits of the training set along the clusters

Fig. 5 ν–analysis SOM structure and results

References

1. Atanassov, K. (2010). On index matrices, part 1: Standard cases. *Advanced Studies in Contemporary Mathematics, 20*(2), 291–302.
2. Atanassov, K. (2010). On index matrices, part 2: Intuitionistic fuzzy case. *Proceedings of the Jangjeon Mathematical Society, 13*(2), 121–126.
3. Atanassov, K., Mavrov, D., & Atanassova, V. (2014). Intercriteria decision making: A new approach for multicriteria decision making, based on index matrices and intuitionistic fuzzy sets. *Issue in IFSs and GNs, 11*, 1–8.
4. Atanassov, K., Szmidt, E., & Kacprzyk, J. (2013). On intuitionistic fuzzy pairs. *Notes on Intuitionistic Fuzzy Sets, 19*(3), 1–13.
5. Atanassova, V. (2015). Interpretation in the intuitionistic fuzzy triangle of the results, obtained by the intercriteria analysis. In *16th World Congress of the International Fuzzy Systems Association (IFSA), 9th Conference of the European Society for Fuzzy Logic and Technology (EUSFLAT)*, Gijon, Spain, pp. 1369–1374.
6. Fidanova, S., Roeva, O., & Paprzycki, M. (2015/2016). InterCriteria analysis of ant colony optimization application to GPS surveying problems. *Issues in IFSs and GNs, 12*, 20–38.
7. Rubio, J. (2020). Stability analysis of the modified Levenberg–Marquardt algorithm for the artificial neural network training. *IEEE Transactions on Neural Networks and Learning Systems, 32*, 3510–3524.
8. Janiesch, C., Zschech, P., & Heinrich, K. (2021). Machine learning and deep learning. *Electron Markets,* 685–695.
9. Morales, E., & Escalante, H. (2022). A brief introduction to supervised, unsupervised, and reinforcement learning. In: *Biosignal Processing and Classification Using Computational Learning and Intelligence, Principles, Algorithms, and Applications* (pp. 111–129).

10. Sotirova, E., Bozov, H., Sotirov, S., Bozova, G., Ribagin, S., & Gonchev, V. (2022). Multicriteria analysis of oncology data during the covid pandemic. In *BioInfoMed: Recent Contributions to Bioinformatics and Biomedical Sciences and Engineering*.
11. Sotirova, E., Stoyanov, V., Sotirov, S., Mirincheva, Z., Bozov, H., & Kostadinov, T. Application of the InterCriteria analysis aproach to a burnout syndrome data. In: K. T. Atanassov, et al. (Eds.), *Uncertainty and imprecision in decision making and decision support—new advances, challenges, and perspectives*. Lecture Notes in Networks and Systems (Vol. 793). Springer.
12. Vassilev, P., Todorova, L., & Andonov, V. (2015). An auxiliary technique for Inter-Criteria Analysis via a three dimensional index matrix. *Notes on Intuitionistic Fuzzy Sets, 21*(2), 71–76.
13. https://ec.europa.eu/eurostat/databrowser/product/page/DEMO_POPPCCTZ

Momentum Adaptation in Convolutional Neural Networks Using a Fuzzy Gravitational Algorithm

Yutzil Poma and Patricia Melin

Abstract This work we propose the adaptation the Momentum parameter, which contributes to the step of updating the parameters of the previous iteration of the stochastic gradient descent used Momentum, which value is scalar between 0 and 1. The value of 0 tells us that there is no contribution from the previous step and 1 indicates that the contribution is maximal. It is proposed to find the best value for the contribution towards the next step since this value influences notably in the recognition of the images. It was demonstrated that the optimization of the parameters with a convolutional neuronal network utilizing fuzzy logic helps obtain the maximize the recognition of the images since this value of the Momentum is not fixed. Therefore, it adapts to the previous step making the downward gradient to be reduced notably and consequently the recognition of the images not only increases but it is also faster since better results are obtained with a decrease in the number of training epochs.

Keywords Optimization · Fuzzy gravitational search algorithm · Convolutional search algorithm · Fuzzy logic · Momentum

1 Introduction

In recent times, science in conjunction with various computational methods have helped us to have more accurate methods to improve the current technology. It has been demonstrated in multiple works that the use of fuzzy logic significantly improves the results for convolutional networks (CNNs), as in [1] where a filtering method is proposed to decrease the image noise using fuzzy logic so that the CNN can detect gender and age with a high accuracy value, or [2] in which recognition is applied to activities in transition with the help of fuzzy logic to separate simple and ambiguous activities, and in [3] they apply the fuzzy control system to a robot that which have classification navigation approach was applied, demonstrating that the

Y. Poma (✉) · P. Melin
Tijuana Institute of Technology, TecNM, Tijuana, México
e-mail: yutpoma@hotmail.com

O. Castillo and P. Melin (eds.), *Intelligent System Design Based on Soft Computing Models*, Studies in Computational Intelligence 1223,
https://doi.org/10.1007/978-3-031-97309-3_8

fuzzy proposal outperformed the robot's own discrete control system. Finally, in [4], "interpretability" oriented layers were added with fuzzy guidelines using a radial basis function (RBF) neural network.

Optimization works using fuzzy logic can also be found, as in [5–7] and in [8], where optimization proposals are made using different types of fuzzy controllers, in this last one they used a type-2 fuzzy logic controller optimized with GSO (galactic swarm optimization) and FA (firefly algorithm) for optimization, as well as in [9] in which the Cuckoo Search algorithm is inspired in the famous parasitism of Cuckoos that is a parasitism of the birds The adaptation of parameters has been of great help to improve the methods, in the case of [10] where the Particle Swarm Optimization also called PSO was utility for adapt the number of layers as well as the number of filters, in addition to the batches that enter for training the convolutional neuronal network. Also, in [11], where the number of filters of each convolution layer in CNN is adjust, in [12], where the Harmony Search Optimization is used for the adaptation of parameters applied to benchmark functions and fuzzy controllers and similarly in the works of [13–15] and [16].

Some works in which the FGSA (Fuzzy Gravitational Search Algorithm) optimization approach was used to adapt to Momentum parameter of neuronal network are present in [16, 17]. In the latter the filter number of convolution layer of a CNN applied to image recognition was searched. In [18] we used parameter for adjust using type-2 fuzzy logic using the FGSA. In [19] and [20], we can observe the application of fractional gradient descent method where they adapt the Momentum for a back propagation neuronal network, in this work a method of Momentum adaptation utilized a fuzzy inference system we proposed, it was applied to a convolutional neuronal network, improving the recognition rates in CNN when the Momentum is static and predetermined to 0.9.

Our support in the present work is the adjust of the momentum in the convolutional neural network, which seeks to adapt the best parameter for the network and the data set using the FGSA method, showing that it does not always have to be 0.9 as it is generally handled in the literature.

The structure for the paper is outlined as such. In Sect. 2 we describe basics concepts, where CNNs are described, as well as the FGSA and topics related to fuzzy logic. In Sect. 3 we can find the proposed Momentum adaptation using a fuzzy inference system. In Sect. 4 we find the results of experimentation with the method proposed in this work. In Sect. 5 we can observe the statistical test, where we verity that our proposal is better. Lastly, in Sect. 6 we present conclusions and future work of our proposed method.

2 Literature Review

At this point, we outline the mains notions for understand the proposed approach for this work.

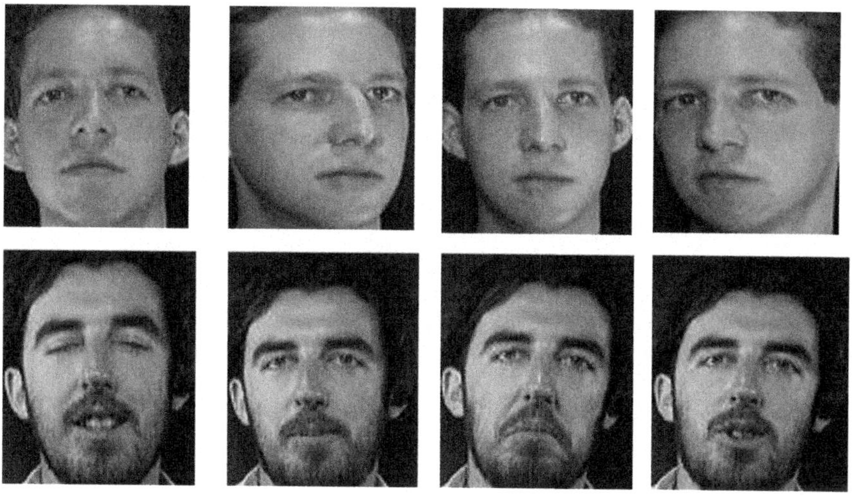

Fig. 1 Some images of the ORL dataset

2.1 ORL Dataset

This database had 40 different images of human faces each with 10 pictures with angles in different positions. The size of 112*92 pixels for each picture and a total of 400 pictures. These images are in grayscale and 70% of the data was used to train the CNN (280 images), while 120 images were used to test the network [21]. In Fig. 1 we observe examples of the ORL dataset.

2.2 Fuzzy Gravitational Search Algorithm (FGSA)

This optimization approach is based on Newton's 2nd law, it's mentions "The change of motion is directly proportional to the magnitude of the driving force and occurs according to the straight line along in which that force is impressed" [22]. The method FGSA is a different version of the GSA (Gravitational search Algorithm) method [23], where there is a implementation of fuzzy inference system is applied to the alpha (α) parameter to adapted then creating the FGSA [24]. The proposed method of FGSA, where the is alpha updated, we can find in Eq. 1.

$$G(t) = G_0 e^{-a\frac{t}{T}} \tag{1}$$

where the letter t represent time and a is the alpha, G_0 is the preliminary value and G is the gravitational constant and alpha.

2.3 Convolutional Neural Network

These deep neural networks are composed of layers that acquire the main features of the images. The initial layer is called convolution layer to extract the image features by way of filters or kernels that have a certain size (for this case study its size is for 3*3 each filter of each convolution layer), then a "feature map" is created and passed to the "pooling" layer. In this layer it reduces the features of the feature map by means of an empty filter taking each maximum value or average of every set of data that are inside "mask", going through the feature map until obtaining a new feature map. Finally, the last layer of the CNN is a classification layer, which is in charge of recognizing the images and/or classifying them [25].

3 Proposed Method

The Momentum is an important part of the network training, as it helps the CNN to find the cost function that is in charge of determining the error between the actual and the estimated value using stochastic gradient descent (SGD).

The convolutional neuronal network that we use in this work has an architecture of two convolution layers and two pooling layers utilizing the ReLU activation function and has one classification layer at the end of this network.

Normally the value of the Momentum is predetermined as 0.9 with a fixed value. When this parameter the closer is to 0 tells us that it does not contribute to the next step or does not do it sufficiently, while when the Momentum is close to 1 it tells us that its contribution is greater to the next step.

The problem with using the Momentum at 0.9 is that it will always use the same value to reduce the error when it reaches the target, but if this parameter is adapted, the time and error is reduced by the time it reaches the target. This proposed method is more accurate when using a fuzzy inference system instead of a static parameter such as momentum at 0.9.

Our proposal consists in adapting the Momentum parameter using the FGSA method for discovering the best aspect for the CNN, we utilizing a fuzzy inference system, which consists of adapter the Momentum and it's consist in one input Er is the Error Network and an output is the Momentum.

The input has 3 membership functions:

EMi which refers to the Minimum Error.

EMe is the Mean Error.

EMax is the Maximum Error.

While the output refers to the Momentum, which has 3 membership functions:

MMi is the Minimum Momentum.

MMe is the Mean Momentum.

MMa is the Maximum Momentum.

This designed fuzzy system is illustrated in Fig. 2.

Below, in Fig. 3 we can find the flowchart of the proposed method.

Three fuzzy rules were designed as follows:

1. If (Er is Emi) then (Momentum is MMa).
2. If (Er is Eme) then (Momentum is MMe).
3. If (Er is EMax) then (Momentum is MMi).

4 Simulation Results and Discussion

The convolutional neural network has been trained with different numbers of epochs, and for each training epoch 30 experiments were performed. Below, in Table I we can find the best experiments for each of the training epochs, together with the average and the standard deviation the performed assessments. In Table 1 we can observe that the highest average was of 96.6666 when the CNN was trained with 40 epochs with a maximum recognition of 95.8333.

Thirty experiments were conducted for each training epoch in which the convolutional neuronal network was maintained with same architecture as in the experiments using the fuzzy inference system. However, the difference is that in this case the Momentum was kept fixed at the default value of 0.9. In Table 2 we can observe the best results obtained from this experimentation, where we can observe that the best value in the average recognition was 84.9167 when the network has been trained with 50 epochs and a maximum average 97.5.

In Table 3 we can find a comparison of the experimentation with our approach method, we use the fuzzy inference system to adapt the Momentum parameter the CNN and experimentation performed when the Momentum is static with a value of 0.9 using the same architecture in both networks.

Our proposed method, consisting in the adaptation of the Momentum parameter, has better results for image recognition, with 94.5556 of average recognition when the network is trained with 40 epochs, while the maximum average recognition with the traditional method using the Momentum fixed at 0.9 has a maximum of 84.9167 when the network is trained with 50 epochs. For this experiment, both networks were trained with 70% of the data.

5 Statistical Test

We performed the statistical test to verify that our proposed method in this work, where we adapt the Momentum of the CNN, compared to the results obtained when the Momentum is static at a default value of 0.9. In this case:

Fig. 2 Proposed fuzzy inference system for adapting Momentum

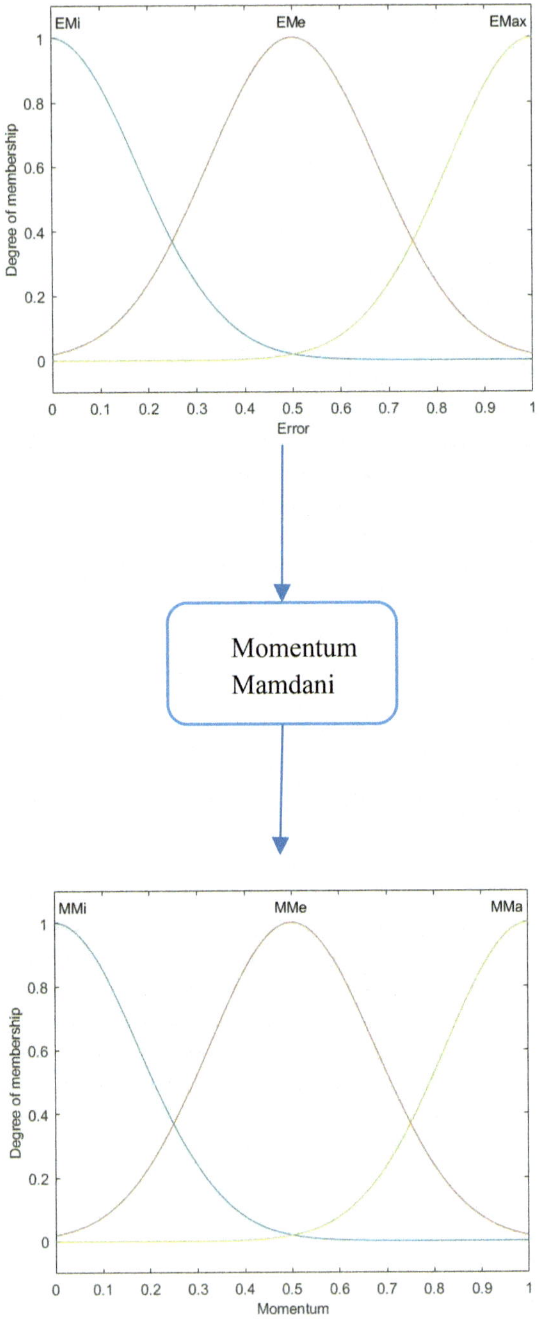

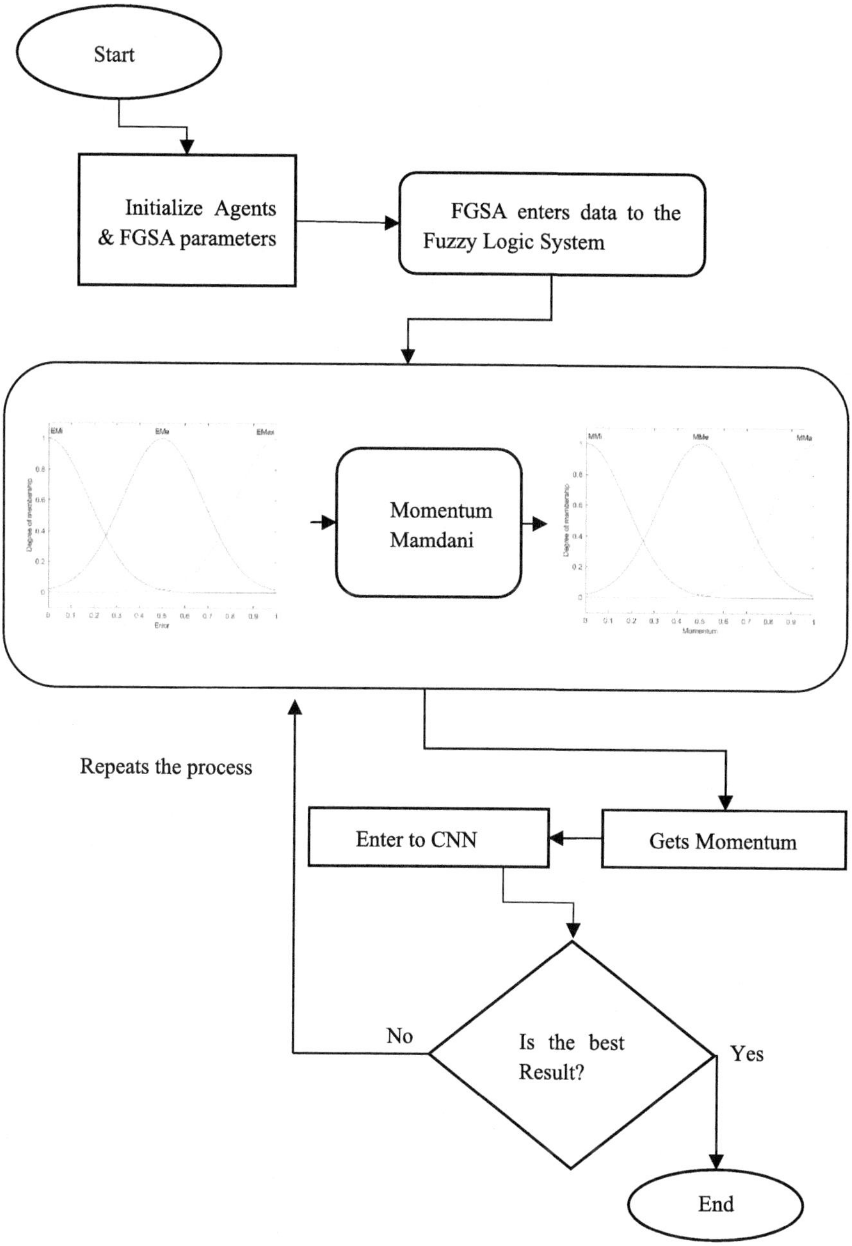

Fig. 3 Flowchart of the proposed method

Table 1 Summary of the best experimental results with proposed method

Epochs	Highest recognition	Average	Standard deviation	Momentum
10	95.83333	92.944	1.06175	0.671185475
20	96.66666	90.576	17.0303	0.671185475
30	95.83333	94.111	0.92675	0.666448913
40	96.66666	94.555	0.86805	0.633327749
50	96.66666	94.250	1.07990	0.666303669

Table 2 Best results of the experiments with momentum 0.9

Epoch	Average	Standard deviation	Max rate recognition
10	81.4722	6.1186	91.6667
20	83.2500	7.2528	94.1667
30	83.8056	5.5566	91.6667
40	82.5278	6.5853	95.8333
50	84.9167	5.9143	97.5000
60	82.9167	6.9987	92.5000
70	82.1389	6.8812	93.3333
80	83.0278	7.1736	93.3333

Ho: RR1 > = RR2 and Ha: RR1 < RR2.

Where RR1 is the variable named the proposed method adaptation of parameter Momentum using Fuzzy Logic in CNN, RR1_Mean is 94.5556 (mean) with Standard Deviation_RR1 is 0.868051.

RR2 is the variable named the momentum in CNN is static with a valor of 0.9, RR2_Mean is 84.9167 (mean) and Standard Deviation_RR2 is 5.9143, in both cases the variables are: significance level is 0.05 and sample sizes are N1 = 30 and N2 = 30.

We observed that $z = 8.832 \geq z_c = -1.645$ $z = 8.832 \geq z_c = -1.645$, we concluded that the null hypothesis is not rejected.

Thus, adapting the Momentum parameter using a fuzzy inference system is better than leaving the Momentum parameter at its default value of 0.9.

6 Conclusions

After the experimentation we concluded that use of a fuzzy inference system for adaptation the Momentum parameter can be valuable. This helps the optimization of the Stochastic Gradient Descent to decrease the cost function of the network, obtains better results and with fewer training epochs than using the Momentum in a predetermined way of 0.9 as it is usually done. Although the use of parameter adaptation with the FGSA method is usually more time consuming, it has been proved

Table 3 Comparison of experiments using the fuzzy inference system to adapt the momentum parameter and when the momentum is static in 0.9 in a CNN

Epochs	Highest recognition–proposed method	Highest recognition–CNN with momentum 0.9	Average–proposed method	Average–with momentum 0.9	Standard deviation–Proposed Method	Standard deviation–with momentum 0.9	Momentum–proposed method	Momentum–with momentum 0.9
10	95.83333	91.6667	92.94444	81.4722	1.061750	6.1186	0.671185	0.9
20	96.66667	94.1667	90.57695	83.2500	17.03039	7.2528	0.671185	0.9
30	95.83333	91.6667	94.11111	83.8056	0.926756	5.5566	0.666449	0.9
40	96.66667	95.8333	94.55556	82.5278	0.868051	6.5853	0.633328	0.9
50	96.66667	97.5000	94.25000	84.9167	1.079904	5.9143	0.666304	0.9

in several works that this method helps to improve results of the convolutional neural network, enhancing the results when used in conjunction with a fuzzy system.

As future work we intend to create an interval type-2 fuzzy and generalized type-2 fuzzy inference systems in order to adapt the Momentum parameter and compare the results, as well as to test the proposal of these fuzzy systems and future work with other databases [26]. Also, to consider the optimization with other metaheuristics [27, 28].

References

1. Tunc, A., Tasdemir, S., Koklu, M., & Cinar, A. C. (2021). Age group and gender classification using convolutional neural networks with a fuzzy logic-based filter method for noise reduction. *Journal of Intelligent & Fuzzy Systems, 42*(1), 491–501. https://doi.org/10.3233/JIFS-219206
2. Kang, J., Kim, J., Lee, S., & Sohn, M. (2020). Transition activity recognition using fuzzy logic and overlapped sliding window-based convolutional neural networks. *The Journal of Supercomputing, 76*(10), 8003–8020. https://doi.org/10.1007/s11227-018-2470-y
3. Born, W., & Lowrance, C. (2018). Smoother robot control from convolutional neuronal networks using fuzzy logic. In *2018 17th IEEE International Conference on Machine Learning and Applications (ICMLA)* (pp. 695–700). IEEE. https://doi.org/10.1109/ICMLA.2018.00110
4. Xi, Z., Panoutsos, G.: Interpretable machine learning: Convolutional neural networks with RBF fuzzy logic classification rules. In *2018 International Conference on Intelligent Systems (IS)* (pp. 448–454). IEEE. https://doi.org/10.1109/IS.2018.8710470
5. Chen, Y., Li, C., & Yang, J. (2023). Design and application of Nagar-Bardini structure based interval type-2 fuzzy logic systems optimized with the combination of backpropagation algorithms and recursive least square algorithms. *Expert Systems with Applications, 211*, Article 118596. https://doi.org/10.1016/j.eswa.2022.118596
6. Calderano, P. H. S., de Mateus Gheorghe, C. R., Teixeira, R. S., Finotti Amaral, R. P., Menezes, I. F. M. (2023). Type-1 and singleton fuzzy logic system binary classifier trained by BFGS optimization method. *Fuzzy Optimization and Decision Making, 22*(1), 149–168. https://doi.org/10.1007/s10700-022-09387-y
7. Carreon-Ortiz, H., Valdez, F., & Castillo, O.: Fuzzy Flower Pollination Algorithm (FFPA): Comparative Study of Type-1 (T1FLS) and interval type-2 Fuzzy Logic System (IT2FLS) in optimization parameter adaptation. *Computación y Sistemas, 26*(2). https://doi.org/10.13053/cys-26-2-4247
8. Bernal, E., Lagunes, M. L., Castillo, O., Soria, J., & Valdez, F. (2021). Optimization of type-2 fuzzy logic controller design using the GSO and FA algorithms. *International Journal of Fuzzy Systems, 23*(1), 42–57. https://doi.org/10.1007/s40815-020-00976-w
9. Guerrero-Luis, M., Valdez, F., & Castillo, O. (2011). *A review on the cuckoo search algorithm, Fuzzy logic hybrid extensions of neuronal and optimization algorithms: theory and applications. Studies in computational intelligence* (Vol. 940, pp. 113–124). Springer. https://doi.org/10.1007/978-3-030-68776-2_7
10. Fregoso, J., Gonzalez, C. I., & Martinez, G. E. (2021). Parameter optimization of a convolutional neuronal network using particle swarm optimization (pp. 149–169). https://doi.org/10.1007/978-3-030-68776-2_9
11. Poma, Y., & Melin, P. (2021). *Estimation of the number of filters in the convolution layers of a convolutional neural network using a fuzzy logic system* (pp. 1–14). https://doi.org/10.1007/978-3-030-68776-2_1
12. Valdez, F., Castillo, O., & Peraza, C. (2020). Fuzzy logic in dynamic parameter adaptation of harmony search optimization for benchmark functions and fuzzy controllers. *International Journal of Fuzzy Systems, 22*(4), 1198–1211. https://doi.org/10.1007/s40815-020-00860-7

13. Zhao, H., Gao, W., Deng, W., & Sun, M. (2018). Study on an adaptive co-evolutionary ACO algorithm for complex optimization problems. *Symmetry (Basel), 10*(4), 104. https://doi.org/10.3390/sym10040104
14. Peraza, C., Valdez, F., Castro, J. R., & Castillo, O. (2018). Fuzzy dynamic parameter adaptation in the harmony search algorithm for the optimization of the ball and beam controller. *Advances in Operations Research, 2018*, 1–16. https://doi.org/10.1155/2018/3092872
15. Castillo, O., Melin, P., Ontiveros, E., Peraza, C., Ochoa, P., Valdez, F., & Soria, J. (2019). A high-speed interval type 2 fuzzy system approach for dynamic parameter adaptation in metaheuristics. *Engineering Applications of Artificial Intelligence, 85*, 666–680. https://doi.org/10.1016/j.engappai.2019.07.020
16. Khatib, M. S., & Atique, M. (2020). FGSA for optimal quality of service based transaction in realtime database systems under different workload condition. *Cluster Comput, 23*(1), 307–319. https://doi.org/10.1007/s10586-019-02923-3
17. Poma, Y., Melin, P., González, C. I., Martínez, G. E. (2020). "Filter size optimization on a convolutional neural network using FGSA. In O. Castillo, P. Melin, and J. Kacprzyk (Eds.), *Intuitionistic and type-2 fuzzy logic enhancements in neural and optimization algorithms: theory and applications. Studies in computational intelligence* (Vol. 862, pp. 391–403). https://doi.org/10.1007/978-3-030-35445-9_29
18. Olivas, F., Valdez, F., Melin, P., Sombra, A., & Castillo, O. (2019). Interval type-2 fuzzy logic for dynamic parameter adaptation in a modified gravitational search algorithm. *Inf Sci (N Y), 476*, 159–175. https://doi.org/10.1016/j.ins.2018.10.025
19. Han, X., & Dong, J. (2023). Applications of fractional gradient descent method with adaptive momentum in BP neuronal networks. *Applied Mathematics and Computation, 448*, Article 127944. https://doi.org/10.1016/j.amc.2023.127944
20. Ren, H.-P., Wang, X., Fan, J.-T., & Kaynak, O. (2019). Fractional order sliding mode control of a pneumatic position servo system. *Journal of the Franklin Institute, 356*(12), 6160–6174. https://doi.org/10.1016/j.jfranklin.2019.05.024
21. Scientific Figure on ResearchGate. (2023). "The ORL Database," Face recognition using ortho diffusion bases
22. González, B., Valdez, F., Melin, P., & Prado-Arechiga, G. (2015). Fuzzy logic in the gravitational search algorithm for the optimization of modular neural networks in pattern recognition. *Expert Systems with Applications, 42*(14), 5839–5847.
23. Rashedi, E., Nezamabadi-pour, H., & Saryazdi, S. (2009). GSA: A gravitational search algorithm. *Inf Sci (N Y), 179*(13), 2232–2248. https://doi.org/10.1016/j.ins.2009.03.004
24. Sombra, A., Valdez, F., Melin, P., & Castillo, O. (2013). A new gravitational search algorithm using fuzzy logic to parameter adaptation. In *2013 IEEE congress on evolutionary computation* (pp. 1068–1074). IEEE. https://doi.org/10.1109/CEC.2013.6557685
25. Kim, P. (2017). MATLAB deep learning: With machine learning. *Neural Networks and Artificial Intelligence.*
26. Ramírez, E., Melin, P., & Prado-Arechiga, G. (2019). Hybrid model based on neural networks, type-1 and type-2 fuzzy systems for 2-lead cardiac arrhythmia classification. *Expert Systems with Applications, 126*, 295–307.
27. Santiago, A., Dorronsoro, B., Nebro, A.J., Durillo, J.J., Castillo, O., Fraire, H.J. (2019). A novel multi-objective evolutionary algorithm with fuzzy logic based adaptive selection of operators: FAME. *Information Sciences, 471*, 233–251.
28. Martínez-Soto, R., Castillo, O., & Aguilar, L. T. (2014). Type-1 and Type-2 fuzzy logic controller design using a Hybrid PSO–GA optimization method. *Information Sciences, 285*, 35–49.

Water Quality Modelling Using Multilayer Neural Network and Multivariate Statistical Techniques

Diana Syulekchieva, Stanislav Popov, Blagovesta Midyurova, Aleksandar Dimitrov, and Sotir Sotirov

Abstract The complexity of water governance makes the involvement of multiple institutions, including citizens, as a prerequisite for the preparation of effective, efficient and appropriate research and policies. Co-creation of knowledge and solutions is increasingly seen as a necessity to address the complex dynamics associated with water pollution. As enabling technologies grow, new approaches that anticipate and assess potential impacts are being developed to foster research and innovation as a framework for responsible research and innovation.

Keywords Aquatic ecosystem · Pollutants · Neural network · Statistical techniques

1 Introduction

Prediction of water quality trends plays an essential role in the field of environmental modeling [1–3], most commonly through linear predictive models (LPMs) and machine learning techniques, pattern and correlation discovery, and anomaly detection in environmental variables [4]. The most widely used method for predicting water quality parameters is through the use of artificial neural networks (ANNs) and their various adaptations [5, 6]. ANN models are able to handle complex linear and nonlinear modeling problems of rivers, lakes, reservoirs, wastewater treatment plants (WWTPs), groundwater, and streams and reliably estimate water quality with appropriate variables. Input variables may include both physical parameters such as temperature and turbidity, chemical factors such as total dissolved solids (TDS) and total nitrogen (TN) [7], or biological such as chlorophyll-a [8]. In a study aiming to evaluate the relationships between water quality parameters is best represented by linear regression, also a predictive model for dissolved oxygen (DO) utilization in a lake is developed using regression analysis and artificial neural network (ANN)

D. Syulekchieva · S. Popov · B. Midyurova · A. Dimitrov (✉) · S. Sotirov
Faculty of Natural Sciance, University Prof. D-R Asen Zlatarov, Burgas, Bulgaria
e-mail: al_dim_2000@abv.bg; adminitrov@uniburgas.bg

O. Castillo and P. Melin (eds.), *Intelligent System Design Based on Soft Computing Models*, Studies in Computational Intelligence 1223,
https://doi.org/10.1007/978-3-031-97309-3_9

techniques [9]. In another study, the authors proposed a deep learning model that integrates deep matrix factorization (DMF) and deep neural networks (DNN) to softly detect and predict BOD_5 values in harbor waters [10]. An artificial neural network (ANN) and support vector model (SVM) were developed to predict total nitrogen (TN) and total phosphorus (TP) concentrations in a river [11]. An ANN model and a three-dimensional circulation model were implemented to predict water temperature in a lake [12]. A model using a back propagation neural network (FFBP NN) was developed to estimate river loads based on selected water quality parameters: biochemical oxygen demand (BOD), chemical oxygen demand (COD), suspended solids (SS) and ammonia nitrogen (NH_3 -N) and a classical statistical multiple linear regression (MLR) model to validate the model [13]. (ANN) were used to estimate the concentration of major ions (Na^+, K^+, $Mg2^+$, $Ca2^+$, HCO^{3-}, $SO_4{}^{2-}$, Cl^- and NO^{3-}) in river water based on pH, alkalinity and temperature [14]. Artificial neural networks (ANN) were used to develop a model to predict the concentrations of DO, BOD, TDS, HCO_3 and SAR index in a river using three different methods namely Principal Component Analysis (PCA), Factor Analysis (FA) and Cluster Analysis (CA) to identify the most influential water quality parameters [15].

1.1 Experiment

The requirements on analytical methods for monitoring these pollutants are laid down in Directive 2009/90/EC [16]. According to this Directive, "The provision of chemical data by Member States must be ensured by analytical methods that comply with relevant international and national standards or analytical methods that are not standardised but provide data of equivalent or better scientific quality. Member States should ensure that all analytical methods used for the purposes of chemical monitoring programmes are validated and documented in accordance with EN ISO/IEC 17,025 or another internationally accepted standard."

Sampling Methods

To determine water quality, samples were subjected to sample preparation and filtration with Sartorius NY 0.45 μm syringe filters. Laboratory analyses were performed with a spectrophotometer DR3900, HACH Lange with cuvettes for the respective analyses: for PO_4-P—cuvettes LCK349, for ammonium—LCK304, for nitrate—LCK339 and for total nitrogen–LCK238.

Physicochemical Analyses of Water

Measurements of physicochemical parameters follow basic physicochemical parameters and standards. The quantification of the physicochemical parameters in the samples analysed shall be carried out by measuring the light absorbance by the UV/VIS spectrometer technique.

For the different water body types, a refined assessment of the physicochemical quality elements and the main indicators for establishing water quality in the selected surface water bodies has been carried out, The parameters pH, electrical conductivity, temperature, and dissolved oxygen/oxygen saturation are measured in situ. All nutrients (NO_3-N, NO_2-N, total N and P-PO_4) are measured in the laboratory according to the specified international standards (analytical methods) - Table 1.

Ammonia nitrogen in water consists of dissociated ammonium and gaseous ammonia. The ratio of the two forms depends on the pH and temperature of the water. Ammonium is primarily determined by a spectrometric method based on the reaction with salicylate and hypochlorite. This reaction produces blue-coloured compounds of the indophenol type whose absorbance is measured at 655 nm. Nitrogen in water can be divided mainly into inorganic nitrogen and organic nitrogen. Typically, inorganic nitrogen includes ammoniacal nitrogen and nitrate nitrogen, in which ammoniacal nitrogen includes free ammoniacal nitrogen (NH_3-N) and ammonium nitrogen (NH_4^+-N), and nitrate nitrogen includes nitrate nitrogen (NO_3-N) and nitrite nitrogen (NO_2-N). The concentration and proportion of different forms of nitrogen in water are often determined by specific circumstances of the water body.

Photometric Determination of Ammonium Ions with Nessler's Reagent [17]

Ammonia reacts in an alkaline medium with potassium mercuriiodide $K_2[HgJ_4]$ to form a yellow–brown precipitate. At low concentrations of ammonia, a colloidal solution suitable for colorimetric determination is obtained. To 50 ml of sample, add 1–2 drops of a solution of the salt of S and homogenise the mixture carefully. Then add 1 ml of Nessler's reagent and homogenise again. After 10 min, measure the absorbance at 400 nm with a DR 3900 spectrophotometer (Hach). The concentration of ammonium ions ρNH^{4+} [$mg.l^{-1}$] was calculated using Eq. 1.

Table 1 Physicochemical parameters and standards

Physicochemical analysis of water samples			
No	Parameters	Standards	Detection limit of the method
1	Determination of active reaction (pH)	BSS* ISO 10523	0.00
2	Determination of electrical conductivity ($\mu S\ cm^{-1}$)	BSS EN 27,888	0.2
3	Determination of dissolved oxygen/oxygen saturation ($mg.l^{-1}$)/(%)	BSS EN ISO 5814	0.02
4	Determination of nitrite nitrogen ($mg.l^{-1}$)	EN 26,777	0.002
5	Determination of nitrate nitrogen ($mg.l^{-1}$)	ISO 7890–1	0.2
6	Determination of total nitrogen ($mg.l^{-1}$)	EN ISO 11905–1	0.5
7	Determination of orthophosphates ($mg.l^{-1}$)	EN ISO 6878	0.01

* BSS–Bulgarian State Standard

$$\rho_{NH4+} = \frac{(A - A_0)V_{max}}{f \times V_s}$$

where: A–the absorbance of the sample;

A_0–the absorbance of the blank sample;

f–the slope in the calibration graph, $L.g^{-1}$;

Vmax–volume of the flask, ml;

Vs–volume of sample in the flask, ml.

Determination of orthophosphates PO_4-P mg/l [18]

After sampling, the sample was filtered through a 0.45 μm syringe filter. The filtrate should have a pH between 3 and 10. A maximum of 40 ml of sample solution is introduced into a 50 ml volumetric flask, 1 ml of ascorbic acid solution and 2 ml of acid molybdate solution are added, brought to the mark with distilled water and stirred. After 10 min, measure the absorbance at 880 nm with a DR 3900 spectrophotometer (Hach).

The orthophosphate concentration ρPO^4 [$mg.l^{-1}$] was calculated using Eq. 2

$$\rho_{NH4} = \frac{(A - A_0)V_{max}}{f \times V_s}$$

where:

A–the Absorbance of the Sample;

A_0–the Absorbance of the Blank Sample;

F–the Slope in the Calibration Graph, $L.g^{-1}$;

Vmax–volume of the Flask, ml;

Vs–volume of Sample in the Flask, ml

Determination of BOD_5, mg/l [19]

The pH of the sample is measured to be within 6.5 ÷ 7.5 and adjusted if necessary. Dilute the sample with distilled water at 20°C. The sample volume corresponds to the expected BOD value. A Lovibond OxiDirect® apparatus is used which operates with ranges from 0 ÷ 40 to 0 ÷ 4000 BOD $mg.l^{-1}$ and sample volumes from 428 ml to 21.7 ml. An anti-nitrification inhibitor is added to the sample and 3 - 4 drops of 45% potassium hydroxide solution (to absorb CO_2) are added to the cap seal. Insert the magnetic stirrer and close tightly. Thus prepared, the sample is annealed at 20°C, then the measurement is started usually for 5 days at the same temperature. The result is expressed in $mg.l^{-1}$, specifying for how many days the measurement was made (5, 10, 20 days).

2 Results and Discussions

Pearson Correlation Coefficient

Correlation analysis is one of the most used methods for measuring the strength or degree of relationship between two or more variables. Pearson's mathematical formulation to quantify the degree of relationship between variables, namely, x and y, can be given as:

$$r_p = \frac{n(\sum xy) - (\sum x)(\sum y)}{\sqrt{\left[n\sum x^2 - (\sum x)^2\right]\left[n\sum y^2 - (\sum y)^2\right]}}$$

where:

n–number of observations;
x–measures of Variable 1;
y–measures of Variable 2;
Σxy–sum of the product of respective variable measures;
Σx–sum of the measures of Variable 1;
Σy–sum of the measures of Variable 2;
Σx^2–sum of squared values of the measures of Variable 1;
Σy^2–sum of squared values of the measures of Variable 2.

Pearson's r returns a value between –1 and + 1. An r value of positive one (+1) indicates a strong positive correlation, while an r value of negative one (–1) indicates a strong negative correlation. An r value of zero indicates no correlation. The degrees of correlation, together with their boundaries are listed in Table 2.

The p-value is the probability of observing a non-zero correlation coefficient in the sample data when the null hypothesis is true. A typical threshold for rejecting the null hypothesis is a p-value of 0.05 or greater. So, if a p-value is less than 0.05, the null hypothesis would be rejected in favour of the alternative hypothesis – that the correlation coefficient is different from zero.

Linear Regression

Linear regression analysis is used to predict the value of a variable based on the value of another variable. The variable you want to predict is called the dependent

Table 2 Degrees of correlation

Degree	Positive	Negative
Perfect	+ 1	–1
High	Between + 0.75 and + 1	Between –0.75 and –1
Moderate	Between + 0.25 and + 0.75	Between –0.25 and –0.75
Low	Between 0 and + 0.25	Between 0 and –0.25
Zero	0	0

variable. The variable you are using to predict the other variable's value is called the independent variable.

This form of analysis estimates the coefficients of the linear equation, involving one or more independent variables that best predict the value of the dependent variable. Linear regression fits a straight line or surface that minimizes the discrepancies between predicted and actual output values. There are simple linear regression calculators that use a "least squares" method to discover the best-fit line for a set of paired data and then estimate the value of X (dependent variable) from Y (independent variable).

Assumptions to be considered for success with linear-regression analysis:

- For each variable: the number of valid cases, mean and standard deviation.
- For each model: regression coefficients, correlation matrix, part and partial correlations, multiple R, R2, adjusted R2, change in R2, standard error of the estimate, analysis-of-variance table, predicted values and residuals, 95-percent-confidence intervals for each regression coefficient,
- Plots: scatterplots, partial plots, histograms and normal probability plots.
- Data: dependent and independent variables should be quantitative.

The correlation heatmap for the combined data for Vaya Lake and Aytoska River shows in Fig. 1.

The result indicates that the best correlation is between BOD5 and Dissolved Oxygen (0.38). The *r*-Pearson for the "BOD5-Temperatire" is -0.17. Both values show decrease in the correlation compared to the results from the first analysis. BOD value is a variable that is dependent on time and temperature. This is because the metabolic rate of aerobic microorganisms depends on temperature and its growth

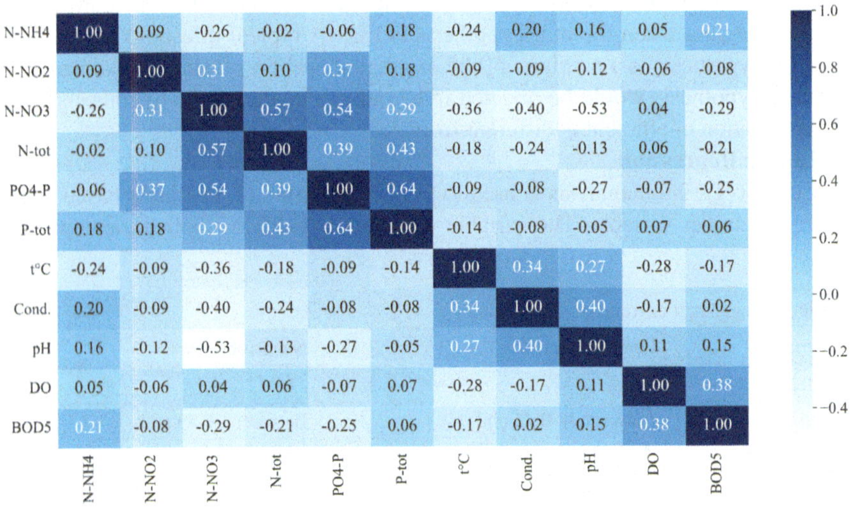

Fig. 1 Correlation heatmap–vaya lake and aytoska river

Table 3 Correlation analysis results

Pair	r-Pearson	p-value
BOD5–DO	0.38	0.003
BOD5–Temperature	–0.17	0.175
BOD5–N–NO_3	–0.29	0.166

continues over time following the microbial growth curve with the exponential and stationary phases. The "BOD5–N-NO_3" pair, however, demonstrates a much higher correlation coefficient (-0.29). Table 3 combines the pairs' r-Pearson value with their p-values.

Based on the comparison between the results from the correlation analysis on both Vaya Lake and Vaya Lake with Aytoska River, we can conclude that relationship between BOD5 and Dissolved oxygen is most reliable and consistent (r-Pearson is over 0.35 and p-values are under 0.05).

From the results obtained, it was found that, there is a direct correlation between some of the significant factors - nitrate content on biological oxygen demand. This is due to the fact that the river Aitoska has a high load along its course, as it passes through many agricultural areas, pastures of animals and the wastewater from the sewage treatment plant of the town of Aitoska is discharged into it Aytos.

Linear regression analysis is performed for the pairs to find if there is any change in the R-statistics for the Vaya Lake indicators and the ones added from Aytoska River – Figs. 2 and 3.

In Table 4 the results from analysis on the different pairs and places of measuremnt are summarized.

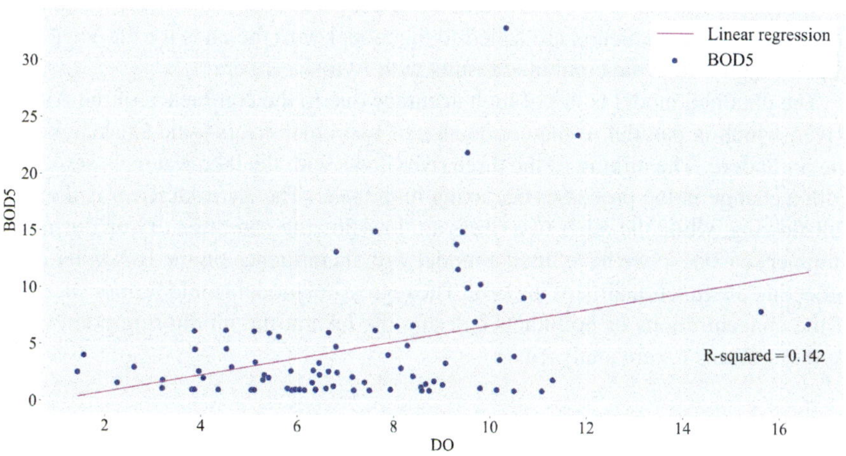

Fig. 2 Linear regression dissolved oxygen-BOD5 (vaya lake and aytoska river)

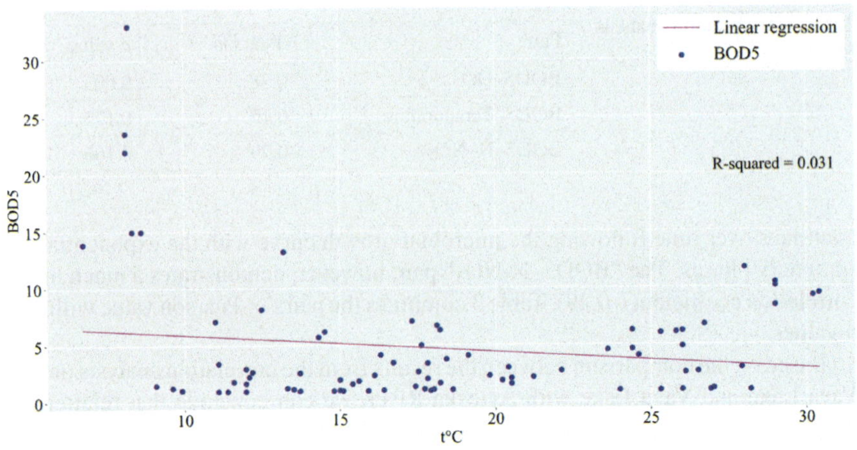

Fig. 3 Linear regression temperature-BOD5 (vaya lake and aytoska river)

Table 4 R-statistics from linear regression analysis

Pair	Place	R-squared
DO-BOD5	Vaya Lake and Aytoska River	0.142
Temperature-BOD5		0.031

Multiple Regression Analysis

For improving the robustness of the model, multiple regression is performed over the collected data (N-NH_4, N-NO_2, N-NO_3, N-tot, PO_4-P, P-tot, Temperature, Conductivity, pH, Dissolved oxygen to BOD5). The results are summarized in Table 5.

Table 5 shows that when the *R*-statistics are substantially improved when the rest of the predictor parameters are added to the model with the ones for the Vaya Lake slightly higher than the combined results with Aytoska River.

The obtained model is not of high accuracy due to the confluence of the Aitoska River, which is parallel to the confluence of two other rivers—the Chakrliyka and the Sarundere. The mixing of the three river flows with the lake waters is associated with a change in the processes occurring in the lake. The Aytoska River is the most full-flowing, while the other two rivers are low-flowing and often dry up during the summer season, so we have fitted a model with its influence on the BOD5 indicator reflecting the functionality of the lake. The regression model would be more accurate if the concentrations of pollutants entering the lake at the monitoring point of the Aytoska-Estuary were analysed.

Table 5 Multiple regression results

	Multiple R	R-square	Adjusted R-square
Vaya Lake and Aytoska River	0.58	0.34	0.25

3 Multilayer Neural Network (Multilayer Perceptron–MLP)

A multilayer neural network (Multilayer Perceptron - MLP) is a type of artificial neural network composed of several layers of neurons. Each neuron in one layer is connected to every neuron in the next layer. Let's look at how a five-layer multilayer neural network works.

Structure of a Six-Layer Multilayer Neural Network.

1. Input Layer: This is the first layer that receives the input data. Each neuron in this layer represents one input feature of the data.
2. Three Hidden Layers: These are the intermediate layers between the input and output layers. They perform nonlinear transformations on the input data through activation functions, which allow the network to learn and model complex dependencies.
3. Output Layer: This is the final layer that generates the network's final output. The output neurons can represent different classes in a classification task or values in a regression task.

Operation of a Multilayer Neural Network.

1. Forward Propagation

    1. The input data {x} = [x_1, x_2, ..., x_n] is fed to the input neurons. The input layer performs no computations and only passes the data to the first hidden layer.
    2. First Hidden Layer–Neurons in the first hidden layer receive a linear combination of the input data, weighted by the connection weights and biases.
    3. Commonly used activation functions include the sigmoid function, ReLU (Rectified Linear Unit), and the hyperbolic tangent function.
    4. Subsequent Hidden Layers - the same process is repeated for all hidden layers. The outputs from each layer serve as inputs for the next layer.
    5. Output Layer - in the final layer, the neurons perform the same computations and pass the results through an activation function suitable for the specific task (e.g., softmax for classification or a linear function for regression).

2. Backward Propagation

    1. Error Calculation
    2. Backpropagation–the error is propagated backward through the network, calculating the gradients of the loss function with respect to the weights and biases.

        - The gradients are calculated using the chain rule, leveraging the gradients from the subsequent layers.

3. Weight Update - after calculating the gradients, the weights and biases are updated using an optimization method (commonly used is stochastic gradient descent SGD or its extensions like Adam):

Our Six-Layer Neural Network

At the input of our neural network, we provide data from the following sources: Ammonium Nitrogen—N-NH4, Nitrite Nitrogen—N-NO2, Nitrate Nitrogen—N-NO3, Total Nitrogen—N-tot, Orthophosphates (as P)—PO4-P, Total Phosphorus (as P)—P-tot, Temperature—t°C, Conductivity, pH—pH, Dissolved Oxygen, Secchi Transparency, Salinity. The output of the neural network is BOD5.

Structure of the Multilayer Neural Network.

- Input layer with 12 inputs (as described above)
- First hidden layer with 122 neurons
- Second hidden layer with 16 neurons (Fig. 4)

- Third hidden layer with 10 neurons
- Fourth hidden layer with 8 neurons
- Output layer with 1 neuron (Fig. 6)

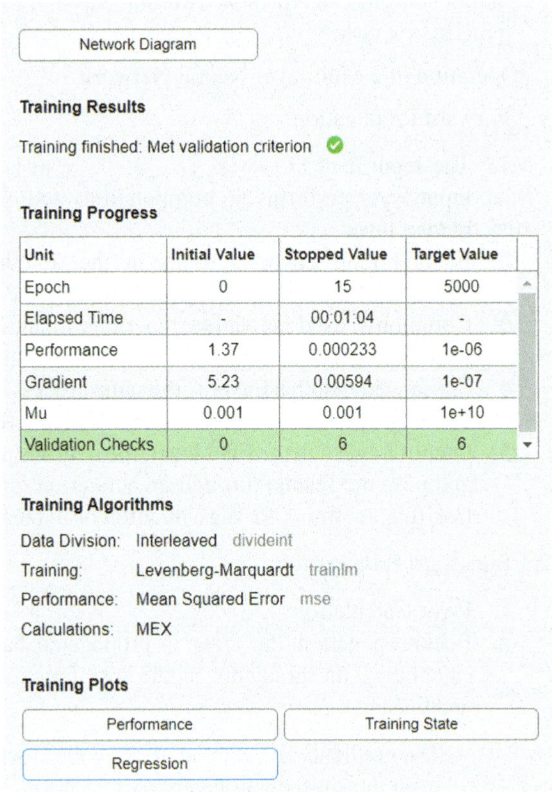

Fig. 4 Structure and learning process of a neural network

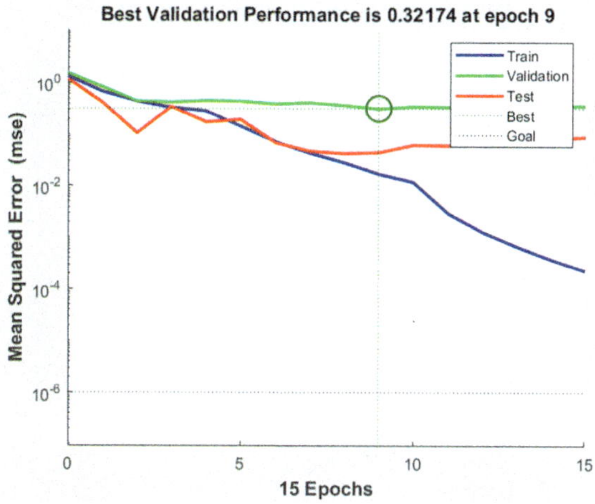

Fig. 5 Neural network training

The training process is depicted in Fig. 5. The results of the training and the corresponding prediction accuracy are of utmost importance. The specific parameters set during the training of the neural network are as follows:

- Total epochs limit: 5000
- Mean squared error (performance): $1*10^{-6}$
- Random split of the training sequence: 70% for training, 20% for testing, and 10% for validation

After training, the average arithmetic error obtained is 1.98%, and the following regression coefficients R are achieved: R for training: 0.98333, R for testing: 0.88261, R for validation: 0.7685, R for the overall training: 0.92204.

4 Conclusions

The increasing complexity of water governance necessitates a multifaceted approach involving diverse institutions and stakeholders, including citizens, to effectively address water pollution challenges. This study highlights the significance of co-creating knowledge and solutions as a vital strategy to tackle the dynamic and intricate nature of aquatic ecosystems. As enabling technologies advance, innovative approaches are emerging, facilitating the anticipation and assessment of potential impacts, thereby fostering a framework for responsible research and innovation.

The research underscores the critical role of predictive models in environmental modeling, particularly in forecasting water quality trends. Artificial Neural Networks (ANNs) have proven to be highly effective in managing complex linear and nonlinear modeling problems across various aquatic environments, including rivers, lakes, reservoirs, wastewater treatment plants, and streams. The integration of physical,

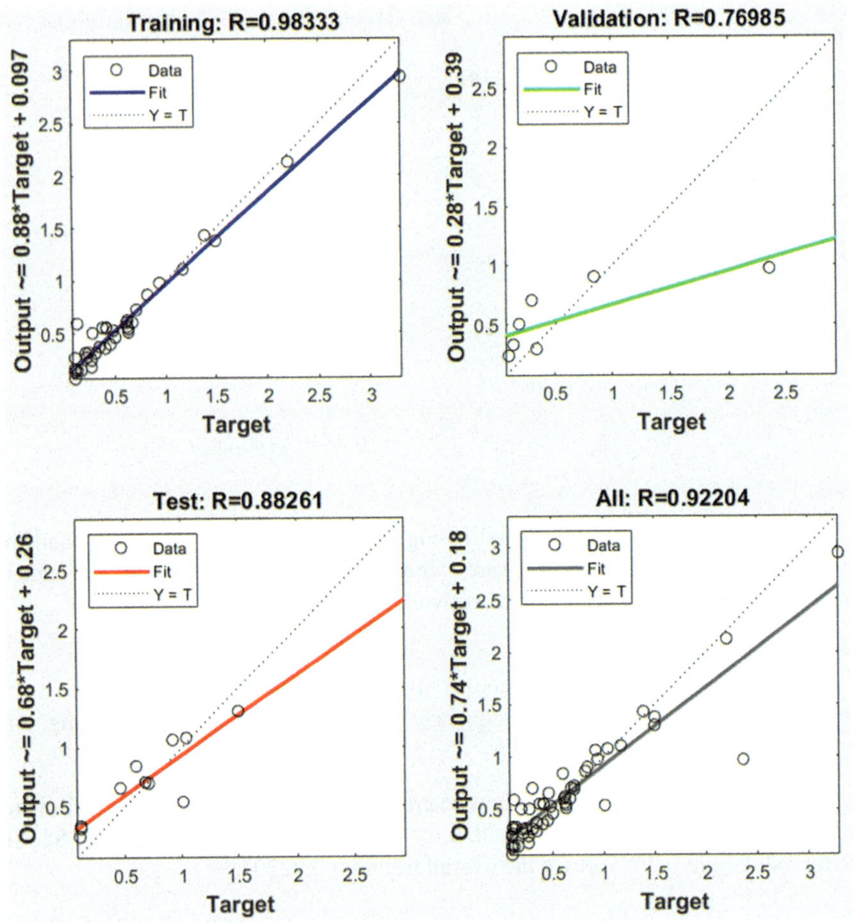

Fig. 6 Regression coefficients of the trained neural network

chemical, and biological parameters in these models enhances their predictive accuracy and reliability.

Various studies have demonstrated the successful application of ANNs in estimating concentrations of key water quality parameters such as dissolved oxygen (DO), biochemical oxygen demand (BOD), total nitrogen (TN), total phosphorus (TP), and other major ions. The utilization of machine learning techniques, including deep learning models, support vector machines, and neural network adaptations, has shown promising results in detecting patterns, correlations, and anomalies in environmental variables.

The findings from this body of research emphasize the necessity for continuous innovation and the adoption of advanced statistical and machine learning techniques in water quality prediction. By leveraging these technologies, researchers and policymakers can develop more effective and efficient strategies to monitor, manage, and mitigate water pollution, ultimately contributing to the sustainable governance of aquatic ecosystems.

References

1. Zhang, Y., Fitch, P., Vilas, M. P., & Thorburn, P. J. (2019). Applying multi-layer artificial neural network and mutual information to the prediction of trends in dissolved oxygen. *Frontiers in Environmental Science, 7*, 46. https://doi.org/10.3389/fenvs.2019.00046
2. Mokarram, M., Pourghasemi, H., & Pham, T. (2024). Enhancing water quality monitoring through the integration of deep learning neural networks and fuzzy method, Marine Pollution Bulletin, Volume 206. *ISSN, 116698*, 0025-326X. https://doi.org/10.1016/j.marpolbul.2024.116698
3. Chen, Y., Song, L., Liu, Y., Yang, L., & Li, D. (2020). A Review of the artificial neural network models for water quality prediction. *Applied Sciences, 10*(17), 5776. https://doi.org/10.3390/app10175776
4. Hino, M., Benami, E., & Brooks, N. (2018). Machine learning for environmental monitoring. *Nat Sustain, 1*, 583–588. https://doi.org/10.1038/s41893-018-0142-9
5. Moeinzadeh, H., Jegakumaran, P., Yong, K., & Withana, A. (2023). Efficient water quality prediction by synthesizing seven heavy metal parameters using deep neural network. *J. Water Process. Eng., 56*, Article 104349. https://doi.org/10.1016/j.jwpe.2023.104349
6. Maier, H. R., & Dandy, G. C. (2000). Neural networks for the prediction and forecasting of water resources variables: A review of modelling issues and applications, environmental modelling & software, 15(1). *ISSN, 101–124*, 1364–8152. https://doi.org/10.1016/S1364-8152(99)00007-9
7. Palabıyık, S., & Akkan, T. (2024). Evaluation of water quality based on artificial intelligence: Performance of multilayer perceptron neural networks and multiple linear regression versus water quality indexes. *Environ., Dev. Sustain.* https://doi.org/10.1007/s10668-024-05075-6
8. Rajaee, T., & Boroumand, A. (2015). Forecasting of chlorophyll-a concentrations in South San Francisco Bay using five different models. *Appl. Ocean. Res., 53*, 208–217. https://doi.org/10.1016/j.apor.2015.09.001
9. Selim, A., Shuvo, S. N. A., Islam, M. M., Moniruzzaman, M., Shah, S., & Ohiduzzaman, M. (2023). Predictive models for dissolved oxygen in an urban lake by regression analysis and artificial neural network. *Total. Environ. Res. Themes, 7*, Article 100066. https://doi.org/10.1016/j.totert.2023.100066
10. Ma, J., Ding, Y., Cheng, J. C. P., Jiang, F., & Xu, Z. (2020). Soft detection of 5-day BOD with sparse matrix in city harbor water using deep learning techniques. *Water Research, 170*, Article 115350. https://doi.org/10.1016/j.watres.2019.115350
11. Liu, M., & Lu, J. (2014). Support vector machine-an alternative to artificial neuron network for water quality forecasting in an agricultural nonpoint source polluted river? *Environ. Sci. Pollut. Res., 21*, 11036–11053. https://doi.org/10.1007/s11356-014-3046-x
12. Liu, W. C., & Chen, W. B. (2012). Prediction of water temperature in a subtropical subalpine lake using an artificial neural network and three-dimensional circulation models. *Comput. Geosci., 45*, 13–25. https://doi.org/10.1016/j.cageo.2012.03.010
13. Khairudin, K., Saufie, Z., Senin, S., Zainudin, Z., & Rashid, A. (2024). Enhancing riverine load prediction of anthropogenic pollutants: Harnessing the potential of feed-forward back-propagation (FFBP) artificial neural network (ANN) models, *Results in Engineering*, 22. *ISSN, 102072*, 2590–1230. https://doi.org/10.1016/j.rineng.2024.102072

14. Nhantumbo, C., Carvalho, F., Uvo, C., Larsson, R., & Larson, M. (2018). Applicability of a processes-based model and artificial neural networks to estimate the concentration of major ions in rivers. *J. Geochem. Explor., 193*, 32–40. https://doi.org/10.1016/j.gexplo.2018.07.003
15. Pany, R., Rath, A., Swain, P.: Water quality assessment for River Mahanadi of Odisha, India using statistical techniques and Artificial Neural Networks.J. Clean. Prod. 417, 137713, ISSN 0959 6526 (2023). https://doi.org/10.1016/j.jclepro.2023.137713
16. Water Framework Directive
17. Water Quality Standard - BDS ISO 7150–1:2002, for the determination of ammonium-nitrogen
18. BSS EN ISO 6878:2005 Water quality. Determination of phosphorus. Spectrometric method with ammonium molybdate (ISO 6878:2004)
19. BSS EN 1899–2:2004 - Determination of biochemical oxygen demand after n days (BODp). Part 2: Method for undiluted samples (ISO 5815:1989, as modified)

A Generalized Net Model of Stochastic Gradient Descent Combined with Dropout Algorithm

Plamena Yovcheva and Vanya Georgieva

Abstract This paper presents a generalized Stochastic Gradient Descent network model in combination with a Dropout method.

Keywords Dropout algorithm · Generalized net · Neural network · Stochastic gradient descent algorithm

1 Introduction

Gradient descent [1] is a prevalent algorithm applied in a wide range of machine learning models, for example logistic regression, graphical model finding, and matrix factorization. In combination with a back-propagation algorithm, this is a standard approach for training artificial neural networks.

According to [2], Stochastic Gradient Descent refers to optimization algorithms and is often used to configure machine learning model parameters. Standard gradient descent uses a gradient to change model parameters. The gradient is typically obtained as the sum of each training element's gradient. At each step, the parameter encoded in vector form, is updated based on the direction of the antigradient. Therefore, to begin utilizing the standard gradient descent algorithm for parameter update, a starting point from a pass through the training data is necessary.

In Stochastic Gradient Descent, an approximation is used in the computation of the value of the gradient as the slope of a loss function calculated for only one training element. Then the parameters get updated in proportion to the approximate slope. In this way, the trainable variables of the model are adjusted following each training object. In the event of large databases consideration, Stochastic Gradient Descent can result in a sizable speed improvement over standard gradient descent.

P. Yovcheva (✉) · V. Georgieva
University of Prof. d-r Assen Zlatarov, Bourgas, Bulgaria
e-mail: plamena.iovcheva@abv.bg

O. Castillo and P. Melin (eds.), *Intelligent System Design Based on Soft Computing Models*, Studies in Computational Intelligence 1223,
https://doi.org/10.1007/978-3-031-97309-3_10

The essence of the algorithm is in the idea of moving steps down. In general, the starting point is chosen at random and the goal is to update that point after every iteration so that the next point is further down the slope.

The speed and results obtained by the algorithm also depend on the chosen value for the learning rate, a flexible parameter. For larger values of the learning rates, the algorithm can take huge jumps along the slope and so may miss the minimum. Therefore, it is generally preferred to choose a low training rate, for example 0.01. It is also possible to demonstrate mathematically that gradient descent can choose larger values for the first part of the descent, if it starts far from the minimum, and then have smaller learning rate close to the minimum. So, this can be a careful way to avoid missing the required result but also obtain it fast enough.

Artificial neural networks commonly have numerous parameters. However, for such machine learning models, congestion is also a great problem. The large size of these neural networks can slow them down, which makes it harder to handle overflow. In this case, the predictions obtained from a large number of different neurons are combined during the test. Dropping out is a technique to deal with this problem [3]. This algorithm can be used in conjunction with other techniques such as Stochastic Gradient Descent.

The weight update in SGD for a given weight w at time t is:

$$w_{t+1} = w_t - \eta \nabla_w \mathcal{L}(w_t, x_i, y_i)$$

where:

- w_t—weight at time t,
- η—learning rate, controlling the step size,
- $\mathcal{L}$—loss function,
- $\nabla_w \mathcal{L}(w_t, x_i, y_i)$ is the gradient of $\mathcal{L}$ w.r.t. w_t, calculated on a single training example (x_i, y_i).

Basically, during training, half the neurons in a certain layer are turned off. The probability that a neuron is turned off during training is a value varying between 0 and 1.

To calculate the weight for the use of each neuron on average (1), where avg_k is the input on average of the weight of a neuron at layer k (containing n neurons); $W_{jk(i)}$ is the weight matrix for iteration i before training starts.

$$avg_k = \frac{1}{n} \sum_{j=1}^{n} \left(\left| W_{jk}^{(i)} \right| \right) \tag{1}$$

In the material, the combined two algorithms—Stochastic Gradient Descent and Dropout method is proposed, and the aim is to borrow the positive part of each of the methods.

1.1 Remarks on Generalized Nets

Generalized nets (GNs) [4–6] are conceptualized very differently which sets them apart from the other versions of the Petri nets. In recent years, GN are used as a tool of choice for the modelling of parallel operations. This has resulted in the development of artificial neural networks [7–12] and data mining methods [2, 13–15].

Firstly, a very large difference between the GNs and the typical Petri nets lies in the "place—transition" connection. The transitions in the GNs case are more complex objects, where each can contain multiple inputs (m in number) and multiple outputs (n in number), where m, n $\geq$ 1.

Every transition is described mathematically as a tuple of seven variables (Fig. 1):

$$Z = < L', L'', t1, t2, r, M, \Box >,$$

where:

(a) L′ and L″ are finite, non-empty sets of places (the input and output places of the transition, respectively). In Fig. 1, for the transition shown, these are:

$$L' = \{l'_1, l'_2, \ldots, l'_m\};$$

and

$$L'' = \{l''_1, l''_2, \ldots, l''_n\};$$

(b) t1 is the current time-moment of the transition's firing;

(c) t2 is the current value of the duration of its active state;

(d) r is the condition of the transition to determine which tokens will pass (or transfer) from the inputs to the outputs of the transition; it has the form of an Index Matrix:

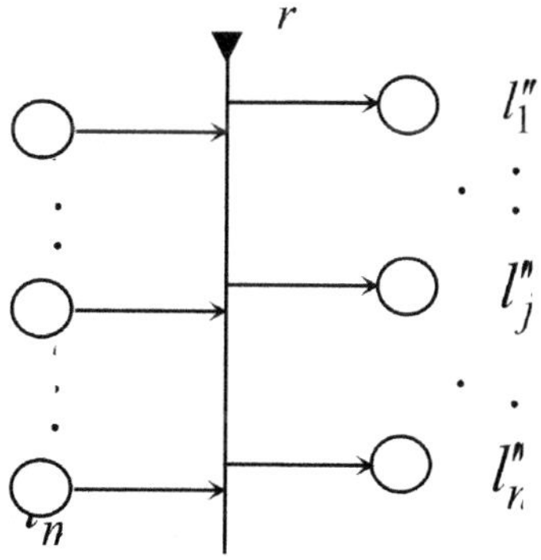

Fig. 1 A GN-transition

$$r = \begin{array}{c|ccccc} & l''_1 & \dots & l''_j & \dots & l''_n \\ \hline l'_1 & & & & & \\ \vdots & \text{ri,j} & & & & \\ l'_i & \text{(ri,j – predicate)} & & & & \\ \vdots & (1 \le i \le m,\ 1 \le j \le n) & & & & \\ l'_m & & & & & \end{array},$$

ri, j is the predicate that corresponds to the i-th input and j-th output place. Given a truth value of the token set to "true", a token from the i-th input place transfers to the j-th output place; an alternative is not possible;

(e) M is an IM of capacities of transition's arcs:

$$M = \begin{array}{c|ccccc} & l''_1 & \dots & l''_j & \dots & l''_n \\ \hline l'_1 & & & & & \\ \vdots & \text{mi,j} & & & & \\ & \text{(mi,j} \ge 0 \text{ – natural number)} & & & & \\ l'_i & (1 \le i \le m,\ 1 \le j \le n) & & & & \\ \vdots & & & & & \\ l'_m & & & & & \end{array};$$

(f) □ is an object of a form like a Boolean expression. The variables in it may be used as symbols serving as labels for a transition's input places, and □ is an expression made from variables and the Boolean connectives ∧ and ∨ and the semantics of which is defined as follows:

$\wedge\ (l_{i1}, l_{i2}, \dots, l_{iu})$—every place $(l_{i1}, l_{i2}, \dots, l_{iu})$ must contain at least one token,
$\vee\ (l_{i1}, l_{i2}, \dots, l_{iu})$—there must be at least one token in all places $(l_{i1}, l_{i2}, \dots, l_{iu})$, where $\{l_{i1}, l_{i2}, \dots, l_{iu}\} \subset L'$.

If the value of a type (calculated as a Boolean expression) is"true", the transition can become active, otherwise it cannot.

2 Generalized Net Model

The networks contain the following cores:

In position S_G—one α_G—core with the specific "Random number generator" for the weight coefficients,

In each position S_F one α_i—core, $1 \le i \le k$, with the specific "Transfer function of the i-th layer of the neural network",

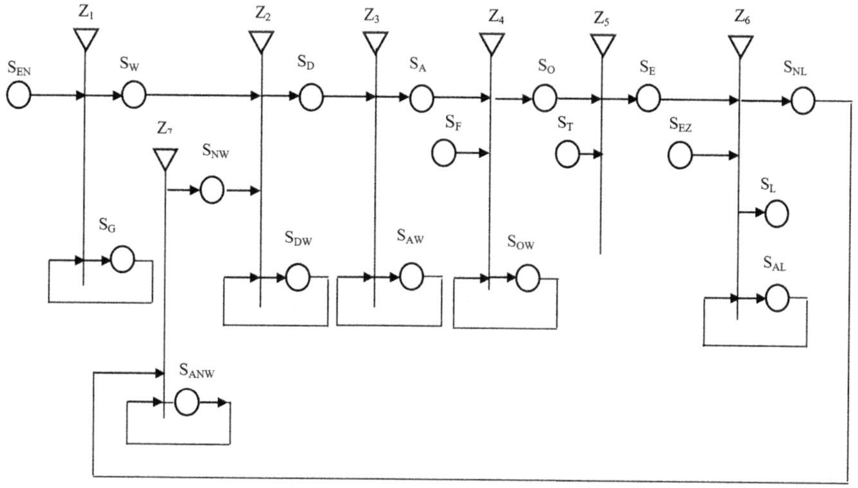

Fig. 2 Generalized network model of Stochastic Gradient Descent combined with Dropout method

In position S_T—one α_t—core with the characteristic "Learning target for the output of a neural network",

In position S_{EZ}—one α_{ez}—core with the specific "Pre-fixed error in the training of a neural network" (Fig. 2).

The generalized network model includes the following transitions:

$$A = \{Z_1, Z_2, Z_3, Z_4, Z_5, Z_6, Z_7\},$$

Z_1—generalization of a random vector for the weight matrix W,

Z_2—calculation of avgK;

Z_3—calculation of the gradient;

Z_4—calculation of the outputs $a_k = F_{K(nk)}$ from the kth layer;

Z_5—determination of the difference between the obtained value (S_O) and the fixed learning target during training and the least square error;

Z_6—determination of whether the neural network is trained or not;

Z_7—calculation of the new weight coefficients;

The generalized network consists of seven transitions with the following description:

The transition Z_1 has the following form:

$$Z_1 = \langle\{S_{EN}, S_G\}, \{S_W, S_G\}, R_1, \vee(S_{EN}, S_G)\rangle,$$

where:

$$R_1 = \begin{array}{c|cc} & S_W & S_G \\ \hline S_{EN} & false & true \\ S_G & W_{G,W} & true \end{array},$$

and

$W_{G,W}$ = "Generated random vector",

At position S_W the kernel receives a specific.

"Weight coefficient W".

The transition Z_2 has the following form:

$$Z_2 = \langle \{S_W, S_{NW}, S_{DW}\}, \{S_D, S_{DW}\}, R_2, \vee(\wedge(S_W, S_{NW}), S_{DW}\rangle,$$

where:

$$R_2 = \begin{array}{c|cc} & S_D & S_{DW} \\ \hline S_W & false & true \\ S_{NW} & false & true \\ S_{DW} & W_{DW,D} & true \end{array},$$

and:

$W_{DW,D}$ = "The calculated average values for W are saved to obtain the outputs of the layers",

In the S_D position the kernel receives a specific.

"average value".

The transition Z_3 has the following form:

$$Z_3 = \langle \{S_D, S_{AW}\}, \{S_A, S_{AW}\}, R_2, \vee(\wedge(S_D), S_{AW}\rangle,$$

where:

$$R_3 = \begin{array}{c|cc} & S_A & S_{AW} \\ \hline S_D & false & true \\ S_{AW} & W_{AW,A} & true \end{array},$$

and

$W_{AW,A}$ = "Save the calculated average values for W to obtain the outputs of the neural network layers",

In the S_A position the kernel receives a specific.

"Output of the neural network with input DN and weight coefficient W".

The transition Z_4 has the following form:

$$Z_4 = \langle\{S_A, S_F, S_{OW}\}, \{S_O, S_{OW}\}, R_4, \vee(\wedge(S_A, S_F), S_{OW})\rangle,$$

where:

$$R_4 = \begin{array}{c|cc} & S_O & S_{OW} \\ \hline S_A & false & true \\ S_F & false & true \\ S_{OW} & W_{OW,O} & true \end{array},$$

and

$W_{OW,O}$ = "The output of the neural network layer is calculated",
In position S_O the kernel receives a specific.

"Output of the neural network with input AN, weight coefficient W and transfer function F".

The transition Z_5 has the following form:

$$Z_5 = \langle\{S_O, S_T\}, \{S_E\}, R_5, \wedge(S_O, S_T)\rangle,$$

where:

$$R_5 = \begin{array}{c|c} & S_E \\ \hline S_O & true \\ S_T & true \end{array},$$

and

In position S_E the kernel receives a specific.

"the value of the least square error during network training".

The transition Z_6 has the following form:

$$Z_6 = \langle\{S_E, S_{EZ}, S_{AL}\}, \{S_{NL}, S_L, S_{AL}\}, R_6, \wedge(S_E, S_{EZ}, S_{AL})\rangle,$$

where:

$$R_6 = \begin{array}{c|ccc} & S_{NL} & S_L & S_{AL} \\ \hline S_E & false & false & true \\ S_{EZ} & false & false & true \\ S_{AL} & W_{AL,NL} & W_{AL,L} & true \end{array},$$

and

$W_{AL,NL}$ = "The neural network is not sufficiently trained",
$W_{AL,L}$ = " The neural network is trained",
In the S_{NL} position the kernel receives a specific.

"the value of the error obtained for recalculating the weight coefficients".

The transition Z_7 has the following form:

$$Z_7 = \langle\{S_{NL}, S_{ANW}\}, \{S_{NW}, S_{ANW}\}, R_7, \wedge(S_{NL}, S_{ANW})\rangle,$$

where:

$$R_7 = \begin{array}{c|cc} & S_{NW} & S_{ANW} \\ \hline S_{NL} & \mathit{false} & \mathit{true} \\ S_{ANW} & W_{ANW,NW} & \mathit{true} \end{array},$$

and

$W_{ANW,NW}$ = "$W_{(n+1)}$ is calculated with the previous values $W_{(n)}$ from the archive.

2.1 Neural Network Training Result

An experiment was conducted with different values for the probability of a given neuron being turned off during the training process. Figure 3 shows a diagram that plots the values of the squared error at a dropout probability of 0.1.

Figure 4 shows a diagram that plots the values of the standard error for a dropout probability of 0.5.

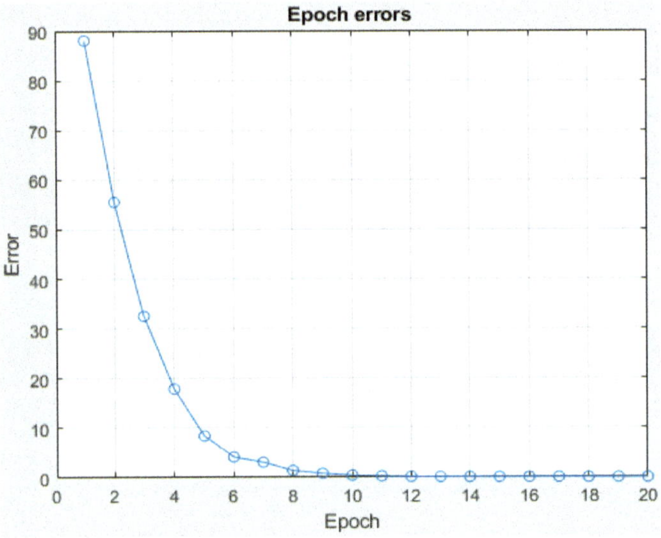

Fig. 3 Values of the squared error at a dropout probability of 0.1

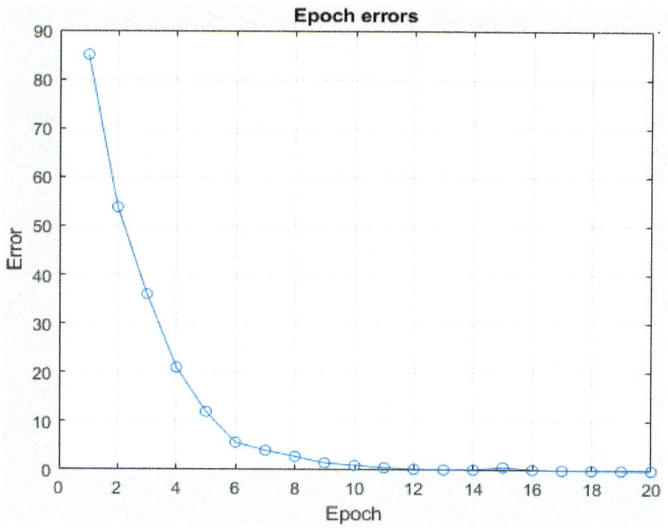

Fig. 4 Values of the standard error for a dropout probability of 0.5

Figure 5 shows a diagram that plots the values of the squared error at a dropout probability of 0.9.

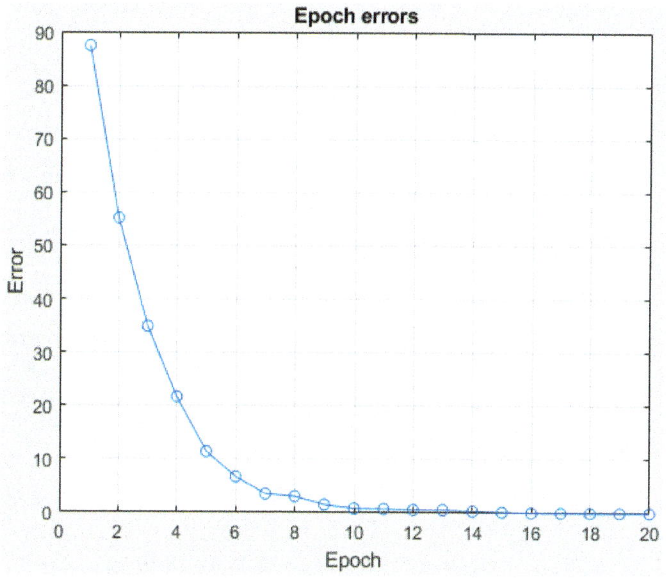

Fig. 5 Values of the squared error at a dropout probability of 0.9

Figure 6 shows a diagram that plots the values of the squared error when applying only stochastic gradient descent (Fig. 7).

The following diagram shows the values of the squared error after each epoch. When using the Stochastic Gradient Descent algorithm without a combination with the Dropout method, we observe higher values of the squared error after each epoch,

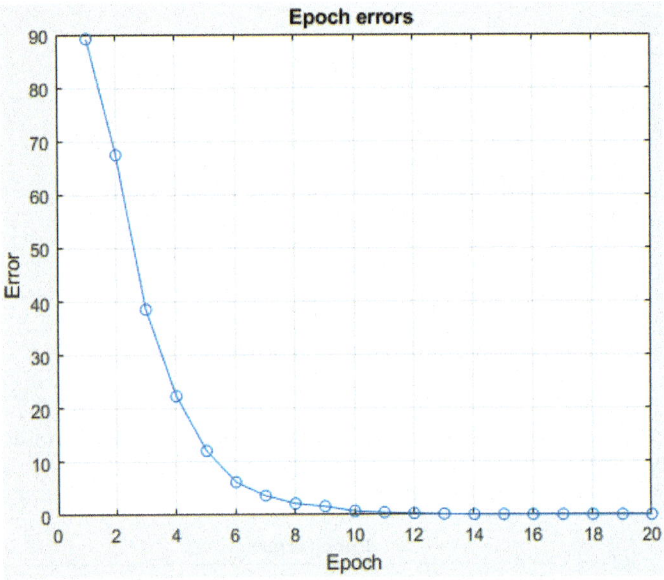

Fig. 6 Values of the squared error when applying only stochastic gradient descent

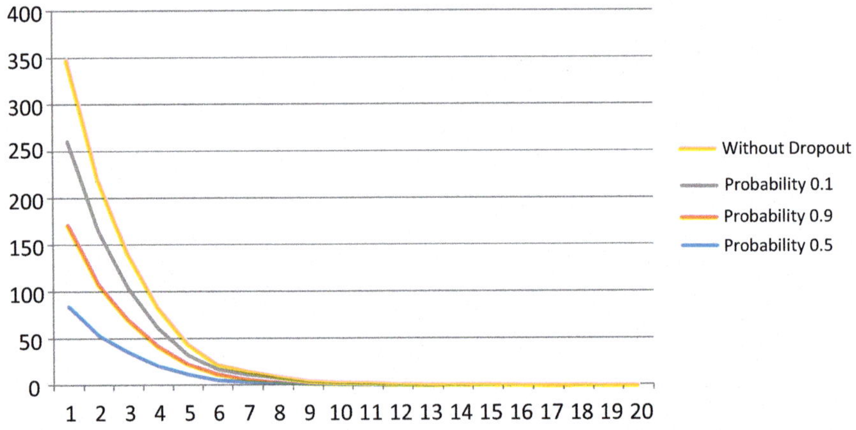

Fig. 7 Values of the squared error when applying only stochastic gradient descent

compared to the other considered cases. The squared error is our criterion for stopping the training.

Two factors mainly influence here. The number of neurons that we need to train in order to be able to perceive the information and the number of neurons that "generate" an error when using it.

With the initial number of neurons, and the corresponding weight coefficients to them, the information is located in all weight coefficients. At the beginning of the training, the error is very large because the number of weight coefficients is large. With a smaller number of weight coefficients, the initial error is smaller. When reducing 10% and 90% of them, the error is similar because when removing the weight coefficients (90%), we also remove the error that accumulates in them. The optimal option is with a dropout probability of 0.5 because the available information manages to "move" into half of the specified number of neurons and, at the same time, remove the error from half of the neurons.

The results of training with the Dropout method with a probability of 0.5 are better than the results of the other experiments, which is why we recommend that you use this probability.

Some of the algorithms in machine learning are not suitable for multi-layer neural networks with many hidden layers, trained with a huge amount of data, since the training itself will be very slow and will require a lot of computing power (??). The algorithm we propose gives us a high learning speed with a large amount of data, since we exclude a neuron.

3 Conclusion

A generalized network describing a process combining the use of Stochastic Gradient Descent and the Dropout method has been created. The results show that the two algorithms successfully achieve their goals, thus, by combining them, various problems can be solved.

The authors are grateful for the support of this work by the European Regional Development Fund through the Operational Program "Science and Education for Smart Growth" under Contract UNITe BG05M2OP001–1.001–0004.

References

1. Yovcheva, P., & Sotirov, S. (2020). A generalized net model of the stochastic gradient descent and dropout algorithm with intuitionistic fuzzy evaluations. *Notes on Intuitionistic Fuzzy Sets, 26*(4), 80–89.
2. Bottou, L. (1998). Online algorithms and stochastic approximations. In: D. Saad, (Ed.) *Online Learning and Neural Networks*. Cambridge: Cambridge University Press.
3. Petkov, T., Jovcheva, P., Tomov, Z., Simeonov, S., Sotirov, S. (2017). A generalized net model of the neocognitron neural network, Lecture Notes in Computer Science (including subseries

Lecture Notes in Artificial Intelligence and Lecture Notes in Bioinformatics), 10333 LNAI, pp. 249–259.
4. Atanassov, K. (2016) Generalized nets as a tool for the modelling of data mining processes, innovative issues in intelligent systems. In *Studies in Computational Intelligence* (vol. 623, pp.161–215).
5. Atanassov, K. (1991). *Generalized nets*. World Scientific.
6. Atanassov, K. (2007). On Generalized Nets Theory, "Prof. Marin Drinov" Academic Publishing House, Sofia.
7. Atanassov, K., Sotirov, S., & Antonov, A. (2007). Generalized net model for parallel optimization of feed-forward neural network. *Advanced studies in Contemporary Mathematics, 15*(1), 109–119.
8. Atanassov, K., Sotirov, S., & Angelova, N. (2020). Intuitionistic fuzzy neural networks with interval valued intuitionistic fuzzy conditions. *Studies in Computational Intelligence, 862*, 99–106.
9. Midyurova, B., Dimitrov, A., Sotirov, S., Petkov, T. (2022) Performance prediction of a microbial fuel cell based on artificial neural networks. *Lecture Notes in Networks and Systems, 374 LNNS*, pp. 202–209.
10. Mustafa, Z., Surchev, S., Milina, R., & Sotirov, S. (2015). A contribution to the recognition of biodiesel fuels according to their fatty acid methyl esters profiles by the artificial neural networks. *Petroleum and Coal, 57*(1), 40–47.
11. Petkov, T., Sotirov, S., & Popov, S. (2018). Generalized net model of optimization of the self-organizing map learning algorithm. *Advances in Intelligent Systems and Computing, 559*, 316–324.
12. Petkov, T., & Sotirov, S. (2013). Generalized net model of the cognitive and neural algorithm for adaptive resonance theory 1. *International Journal Bioautomation, 17*(4), 207–216.
13. Atanassov, K., Sotirov, S. (2020). Generalized nets as tools for modelling of the neural networks. In *2020 IEEE 10th International Conference on Intelligent Systems, IS 2020—Proceedings*, art. no. 9199965, pp. 521–525.
14. Kralev, K., Mirincheva, Z., Sotirov, S. (2024). Hybrid convolutional neural network with intuitionistic fuzzy estimations for detection of kidney damage in patients with diabetes mellitus, *Lecture Notes in Networks and Systems, 1088 LNNS*, pp. 497–502.
15. Kralev, K., Sotirov, S., Sotirova, E., Petrova-Geretto, E., Petkov, D. (2024). Modular convolutional neural network for children pneumonia detection via x-ray images. *Lecture Notes in Networks and Systems, 1088 LNNS*, pp. 319–323.
16. Kralev, K., Mirinchev, N., Sotirov, S., Sotirova, E., Cholakova, Z. (2024). A modular deep convolutional neural network for imroving accuracy in prostate biopsies. *Lecture Notes in Networks and Systems, 1088 LNNS*, pp. 302–308. Author, F., Author, S., Author, T. (1999) Book title. 2nd edn. Publisher, Location.
17. Krawczak, M., Sotirov, S., & Sotirova, E. (2017). Modeling parallel optimization of the early stopping method of multilayer perceptron. *Studies in Computational Intelligence, 657*, 103–113.
18. Petkov, T., Sotirov, S. (2015). Describing and simulation of the generalized net model of the ART2 neural network. Comptes Rendus de L'Academie Bulgare des Sciences, *68*(11), 1431–1438.
19. Petkov, T., Sotirov, S., & Surchev, S. (2016). Generalized net model of person recognition using ART2 neural network and viola-jones algorithm. *Advances in Intelligent Systems and Computing, 401*, 257–265.
20. Srivastava, N., Hinton, G., Krizhevsky, A., Sutskever, I., Salakhutdinov, R. (2014). Dropout: a simple way to prevent neural networks from overfitting. *The Journal of Machine Learning Research, 15*(1), 1929–1958.
21. Sotirov, S., Sotirova, E., Melin, P., Castillo, O., & Atanassov, K. (2016). Modular neural network preprocessing procedure with intuitionistic fuzzy InterCriteria analysis method. *Advances in Intelligent Systems and Computing, 400*, 175–186.

22. Sotirov, S., Sotirova, E., Surchev, S., Petkov, T., Georgieva, V. (2021) Generalized net model of the deep convolutional neural network. *Advances in Intelligent Systems and Computing, 1081 AISC*, pp. 131–138.
23. Sotirov, S., Petrova, Y., Bozov, H., Sotirova, E. (2022) A hybrid algorithm for multilayer perceptron design with intuitionistic fuzzy logic using malignant melanoma disease data. *Lecture Notes in Networks and Systems, 504 LNNS*, pp. 665–672.
24. Sotirova, E., Stoyanov, V., Sotirov, S., Mirincheva, Z., Bozov, H., Kostadinov, T. (2023). Application of the intercriteria analysis aproach to a burnout syndrome data. *Lecture Notes in Networks and Systems, 793 LNNS*, pp. 227–236.
25. European Regional Development Fund through the Operational Program "Science and Education for Smart Growth" under Contract UNITe BG05M2OP001–1.001–0004
26. Tsuruoka, Y., Tsujii, J., Ananiadou, S. (2009). Stochastic gradient descent training for l1regularized log-linear models with cumulative penalty. In *Proceedings of the AFNLP/ACL '09.*

Dropout Prediction for Higher Education: Data Sets and Methods, a Brief Overview

Julia Nolazco Nuñez and Mario García-Valdez

Abstract Dropout prediction represents one of the head educational objectives in higher education. One problem with making this prediction is generating or acquiring the data sets to train and validate the models that will make it. As a preliminary approach to these challenges, this study presents a comparison of five datasets used by researchers in projects related to dropout prediction, along with a summary of machine learning techniques such as decision trees, random forests, linear regression, artificial neural networks, support vector machine, random forests, among others. Finally, this investigation will provide to researchers and practitioners of educational data mining useful and important information to achieve dropout prediction goals effectively and efficiently, highlighting relevant features and particularities of datasets and what are the most popular techniques to find insights for dropout prediction.

Keywords Machine learning · Educational data mining · Prediction · Classification · Dropout

1 Introduction

From a general perspective, well-educated people provide more benefits to their countries, which is vital for increasing economic progress and employment rates. Failure in higher education could represent an obstacle, impacting students' lives and all the elements involved [22]. In Latin America, half of all the students leave school earlier, and only half of those remaining students graduate on time [14, 21]. As a result, higher education institutions need to identify what are the key factors

J. N. Nuñez (✉) · M. García-Valdez
TecNM, Tijuana Institute of Technology, Tijuana, Baja California, México
e-mail: jenn13054@gmail.com

M. García-Valdez
e-mail: mario@tectijuana.edu.mx

O. Castillo and P. Melin (eds.), *Intelligent System Design Based on Soft Computing Models*, Studies in Computational Intelligence 1223,
https://doi.org/10.1007/978-3-031-97309-3_11

contributing to this situation and implement measures to reduce it, enhancing student performance.

One of the main objectives established in the National Development Plan 2023–2024 is to promote excellence training through innovation and the comprehensive and continuous improvement of higher education [10]. According to The Official Daily of the Federation (DOF), in Mexico, there are high levels of school leaving rates, presenting 8.8% at the national level, variable behavior depending on the federal entity, for example Baja California, Colima, Veracruz, Quintana Roo, and Sinaloa shows rates above 13.5%.

Using new computational techniques, such as Machine Learning (ML), facilitates the identification of factors impacting this problem. Multiple studies have obtained remarkable levels of precision, higher than 80% [12], serving as a tool to facilitate the academic process, taking assertive decisions and applying strategies at an institutional level. In the educational field, data analysis needs to previously consider some aspects, such as the hybrid nature of the information and the large amount of data generated by various stakeholders, such as teachers, students, and their interactions, administrative information, and mobile platforms, among others.

There are not many standard articles or datasets that researchers can use to validate the data mining techniques they propose. These datasets are usually expensive and time-consuming to produce, which makes them an asset for the research community. The problem is that sometimes, published datasets need to be better documented, have a user-friendly format, have many missing values, and have other issues that complicate their adoption and the definition of a benchmark that widely contributes to the evaluation, new experimentation, and support the improvement of machine learning models.

Based on a systematic literature review of school dropout prediction articles and ML methods, we centered on five datasets to compare them identifying their main facts.

They were selected based on their popularity in the various articles, the number of citations, and the results obtained from their use. In addition, we chose datasets that presented diversity in their attributes and the number of records that compose them.

Another point of relevance for new data mining studies in the educational area is to understand in advance the current trends in the use of advanced machine learning techniques which focus is predicting student performance and identifying patterns. This article summarizes these techniques, and the results obtained using them, giving a broader overview of the beginning of new developments and research.

The present article follows the next organization: Sect. 2 describes an overview of machine learning techniques common in data mining analysis. Section 3 compares five datasets used in dropout prediction and similar problems. Section 4 summarizes the comparative analysis of datasets and the different ML techniques. Finally, in Sect. 5 conclusion is shown as well as future work.

2 Overview Machine Learning Techniques

The most frequent machine learning techniques employed in the educational field [4, 5, 11, 15] are outlined below, classified according to the data mining task for which they are applied.

2.1 Classification

Classification methods are applied for predictive purposes, by classifying a dataset through with a model into numerous classes. The process for classification often includes two stages: first, generated a classifier model, and second, using this classifier to label new data items without the unknowing of class labels. The primary purpose is to predict some future values of the data with only the significant features, splitting them into preset classes. A brief outline of different classification techniques is provided below.

2.1.1 Decision Trees

This technique is based on conditions that build a tree-like diagram. The process of building a decision tree is simple, it has as an input a set of features and gives class labels as output. The first node, called "root" is placed on top of the tree, branching out, each branch describes a condition further linked with the next node. Using an if then algorithm, the process of construction continues until reaching a terminal node or "leaf", which is labeled as a class or decision, this approach is better known as a divide-and-conquer manner [21]. This technique is the most straightforward, and thus, it is very easy to understand how it works. Some of the most popular algorithms for this technique are: ID3, C4.5, C5.0, etc. These recursive algorithms basically seek to divide the data set into increasingly homogeneous or pure sub-groups. Figure 1 presents a simple decision tree model, which describes a prediction for students' results.

2.1.2 Artificial Neural Network

It is a computational technique inspired by the function of the human brain. Artificial Neural Network (ANN) is designed to recognize patterns, learn from data, and make predictions. It could be used to solve data mining problems. An ANN comprises interconnected layers of nodes (neurons) and directed networks between them divided into three layers: the input layer, which receives the raw data; the hidden or process layer; and the output layer, where the final result is produced. Figure 2 describes the workings of a basic ANN model.

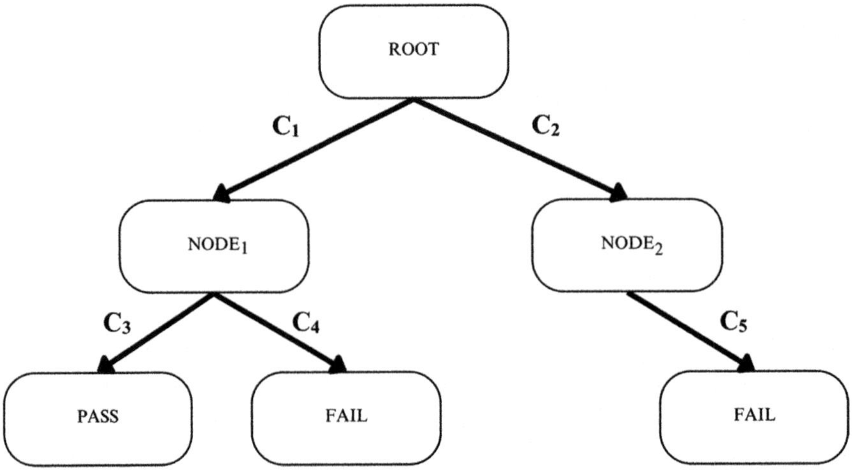

Fig. 1 Decision tree (DT) model [4]

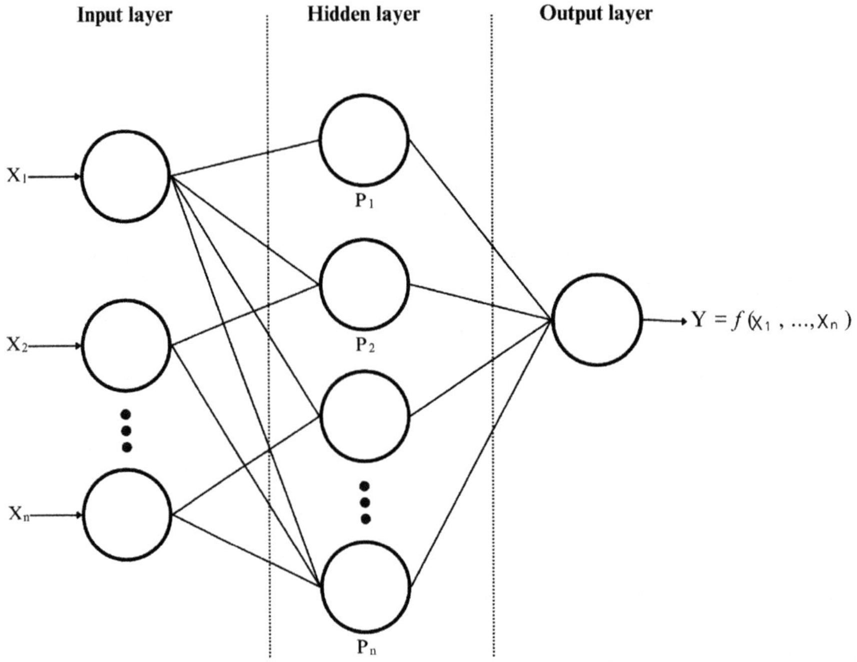

Fig. 2 Artificial neural network

The prediction output depends on the training of the ANN model, the types of data and the amount of it, as well as the data pre-processing are significant steps that help to enhance results.

2.1.3 Bayesian Classifiers

There are a few types of Bayesian classifiers. A Naïve Bayes (NB) Classifier assumes "naively" the conditional independence, given the class of the features; this kind of assumption is rarely found in real data but makes it very efficient to compute. It predicts the category or class of a given data point by calculating the probability that the data belongs to each class and gives the features of that data [20].

Bayes' theorem describes the relationship between conditional probabilities:

$$P(A|B) = \frac{P(B|A) \cdot P(A)}{P(B)}$$

where P(A|B) is the posterior probability of event AAA (class) given that BBB (feature) has occurred. P(B|A) is the probability of observing BBB given that AAA is real or true. The prior probability, defined as P(A) is the probability of AAA before observing any evidence and the total probability is given for P(B).

NB has different approaches, depending on the type of data, e.g. for continuous data, Gaussian type is proper, because the model assumes that the features follow a normal distribution or Gaussian. For discrete data, the multinomial Naïve Bayes, can be applied in text classification problems, and finally Bernoulli Naïve Bayes that works with binary features, useful for tasks like text classification where presence or absence of a word matters.

They are helpful, simple, and fast in training and prediction tasks (e.g., spam detection, document categorization, sentiment analysis, medical diagnosis).

2.1.4 Support Vector Machine

Supervised learning algorithm, that uses labeled classes for different tasks (classification, regression, prediction). Support Vector Machine (SVM) divides the dataset belonging to different classes (e.g., "pass" vs. "fail"), it plots instances on a 2 or 3-dimensional space and draws the optimal boundary (hyperplane) between different classes. This decision boundary divides the data into different classes; and the model refers to it in different ways, depending of the dimensions of the space: a line, a plane and a generalized plane. Figure 3 represents general SVM model for student classification, divided by a hyperplane into two classes, pass/fail corresponding to final student performance, where the elements that fall on one side of the red line (hyperplane) are considered to belong to one class; the nearest is called a vector (x_1, x_2) [12]. The different classes are clearly represented with a wider separation of data.

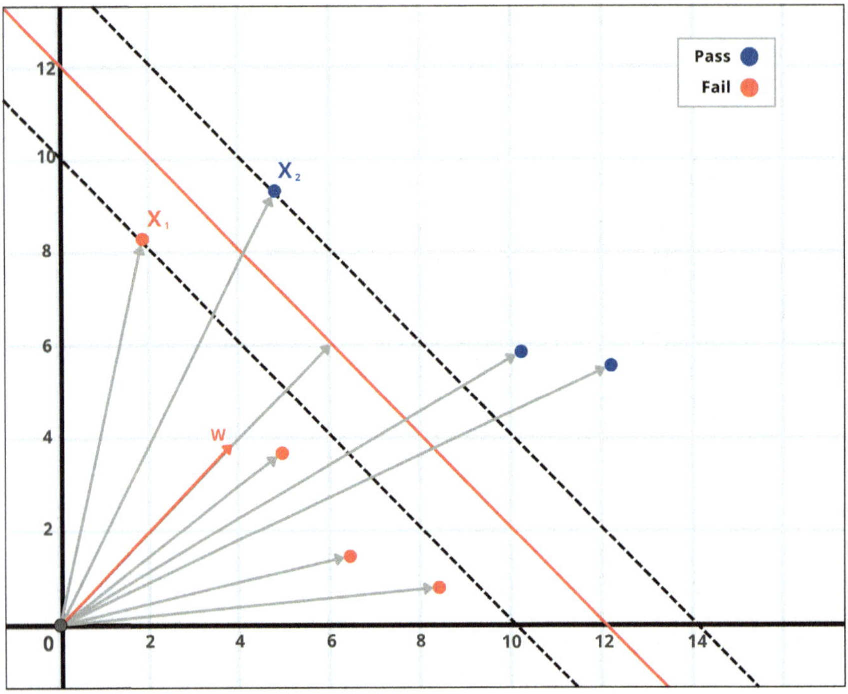

Fig. 3 Support vector machine [20]

2.2 Clustering

Cluster analysis refers to techniques used to group data based on the similarity to each data point, this set of groups are called "clusters". Notably, this method can work without a category definition or class label, this approach is better known as unsupervised learning.

There are a few possibilities for cluster algorithms that can be implemented under this kind of learning, for example k-means, hierarchical clustering, Gaussian mixture, and DBSCAN.

2.2.1 Random Forests

This is an ensemble machine learning algorithm; as stated by Breinman [8], it is defined literally as: "a classifier consisting of a collection of tree-structured classifiers $\{h(x, \Theta_k), k = 1, \ldots\}$ where the $\{\Theta_k\}$ are independent identically distributed random vectors and each tree $h(x, \Theta_k)$ casts a unit vote for the most popular class at input x."

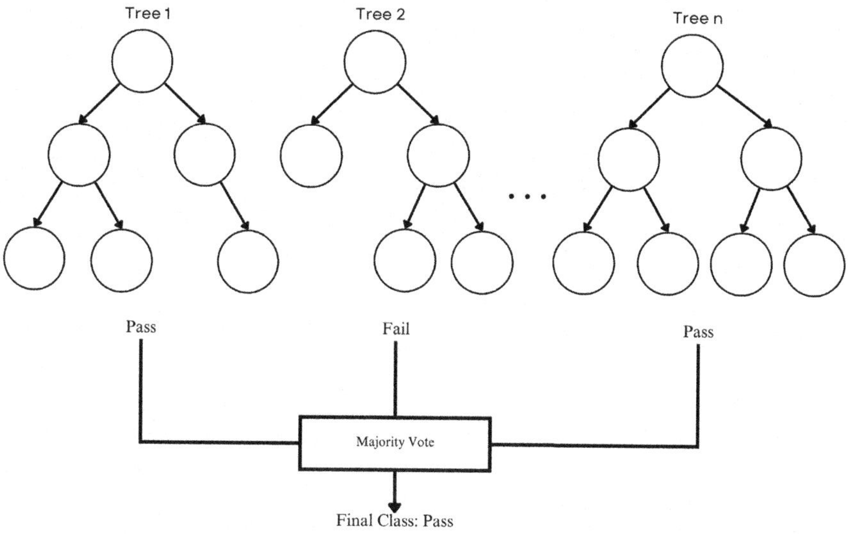

Fig. 4 Random forests (RF) [5]

It's considered an improvement of the traditional decision trees, combining the predictions made by multiple decision trees, each of them is generated from the values of an independent set of random vectors, and the prediction results of all are merged to achieve the final outcomes, which is more accurate. Figure 4 shows the working of this classification model, for a pass/fail result in students' evaluation.

2.3 *Prediction*

Consist in the use of past data to anticipate and make predictions of the future; it has been defined by Han [12] as the task of predicting continuous or ordered values for a given input. Regression is one of the most applied prediction techniques.

Regression analysis could model the relationship between one or more independent variables and a dependent variable. Specifically, in education, this technique aims to identify which students are most likely to succeed, which need remediation in a subject, and which are susceptible to failing in their classes and dropping out.

2.3.1 Linear Regression

Statistic method, naturally work for classification in contexts where both attributes and outcomes are numerical. The core idea of this method is to represent the class as a linear combination of the features, as follows:

$$x = w_0 + w_1 a_1 + w_2 a_2 + \ldots + w_k a_k$$

where all the w_1, w_2, ..., w_k attributes have predetermined weights, x is the class, a_1, a_2, ..., a_k are the attributes values. Linearity could be a disadvantage for this method, for data that presents a non-linear dependency, this line will not fit for all the features, though this linear model can be a foundation for developing more complex methods [21].

3 Datasets Description for Data Mining Analysis

Comprehension of context is essential to jump into an assertive data analysis. Based on this, is presented a description of five public datasets used specifically to predict dropout, performance, and success in higher education students. Selected based on their popularity in the various articles, the number of citations, and the results obtained from their use. In addition, we chose datasets that presented diversity in their attributes and the number of records that compose them.

The first dataset is found in a paper by Realinho and Machado [19]; in this work they predict students' dropout and academic success with a publicly available dataset that has 4,424 instances and 36 features referring to academical data, socio and macroeconomic, demographic information, and performance. Created from the Polytechnic Institute of Portalegre, records belong to students registered from diverse disciplines, in the years between 2008/2009 and 2018/2019. The data helps build models to predict dropout and academic success in a specific time, in this casa at the end of the first and second semesters. Data is classified under: dropout, enrolled and graduated. The author suggests using this dataset as a benchmarking reference for similar problems and machine learning areas.

Another dataset is found in the work by Yilmaz & Sekeroglu [22], through artificial intelligence techniques, performed a higher education students performance evaluation analysis according to the final grades of students from the Atatürk Faculty of Education and Faculty of Engineering of Near East University data, Turkey in 2019. Each record is a list of answers to a survey provided by students, divided into personal questions, family, and education habits. As a result, a dataset with 145 instances and 31 attributes was created. The primary purpose of it is to predict students' end-of-term.

Furthermore, Kuzilek et al. [13] define the well-documented dataset OULAD (Open University Learning Analytics Dataset), compiled from courses presented at the Open University (OU) in the academic years 2013–2014 and shows demographic information in conjunction with Virtual Learning Environment (VLE) students' interactions. This set of data presents information about 22 courses, with 32,593 instances, divided into seven CSV files with information about assessment results, interaction logs, student data, and information courses. The dataset "Open University Learning Analytics dataset" is available under a CC-BY 4.0 (Creative Common) license by

Jakub Kuzilek and their source in the UCI machine learning repository. It is continuously evaluated to be up to date with the actual student population. The main applications of this data are to predict academic performance, student engagement, and early detection of at-risk students. In their work, they use several machine learning algorithms to address these tasks. A limitation of this dataset is that it presents messy data and requires tough preprocessing and merging tables, facts that may be considered for its use.

The dataset High School Student Performance & Demographics has been described in Cortez & Silva [8] as data collected from two public high schools in Portugal, in the academic years 2005–2006. The main data sources to collect information were mark reports and questionnaires, complemented with students' information, social, parents' information, and school-related features; this data is presented in two CSV files with 395 instances related to Mathematics and Portuguese courses. The study predicts, through 34 attributes, the students' success. Another use is for identification of key attributes that influence educational success/failure through two core goals: binary classification (fail/pass), five-level classification (I excellent – V insufficient), and regression values from zero to twenty (0–100%).

The last one is the dataset defined by Delahoz [9], which presents information about 12,411 engineering students, with 44 attributes regarding academic data and social and economic information. It evaluates the academic performance of students at the end of high school through the SABER 11 test and in the final year of their professional training applying the SABER PRO evaluation.

The original data presents a complete panorama of demographic variables, such as internet, TV, computer, phone, mobile, and similar, but requires some preprocessing work. It is accessible through the Mendeley data repository as a spreadsheet formatted in an Excel file (xlsx).

All the previously mentioned datasets show various information related to students differently. It is important to identify the primary study objective to make an assertive choice of benchmark to make comparisons and start the validation of models generated.

A summary of the datasets is presented in Table 1. Enlist the main characteristics for each dataset: name, number of instances, attributes, format, and if they're useful for classification or not.

4 Comparative Analysis: Datasets and Machine Learning Techniques

This section presents a comparative analysis of dataset particularities, and the different Machine Learning techniques that are applied to develop educational solutions.

Education and Data Mining (DM) are two fields that work together through Educational Data Mining (EDM), working specifically with data extracted from scholars

Table 1 Dataset summary

Id	Name	Instances	Attributes	Format	Class
A	Predict students' dropout and academic success	4,424	36	csv	X
B	Higher education students performance evaluation	145	31	csv	X
C	Open university learning analytics dataset	32,594	12	csv	
D	High school student performance & demographics	395	34	csv	
E	Data of academic performance evolution for engineering students	12,411	44	xlsx	

and their context, identifying patterns and developing methods to make discoveries in pro to understand students and their particularities [5].

All the datasets mentioned in previous content have been used in several studies to achieve purposes like predicting dropout, success, or performance in higher education students. The following table, Table 2, resumes the dataset ids and techniques that investigators use to reach their goals. We can observe that the initial preferences are for the decision trees technique rather than others, but they work in conjunction with more complex methods like Radial Basis Function Neural Network (BPNN) and Back Propagation Neural Network (RBFNN).

From the point of view of research and the development of solutions, he dataset's size and the model's complexity are essential characteristics for the quality of the results generated in the evaluation of the models. The greater the number of characteristics, the less unstable the results.

The more relevant features for this data analysis are focused on academic or performance aspects, demographic information, and student engagement. Considering these features and providing only the most relevant and useful ones is essential.

Table 2 Techniques/Datasets

Dataset Id	DT	SVM	ANN	RF	LR	Other
A	X	X		X	X	
B	X		X		X	BPNN, RBFNN
C	X	X		X		KNN
D	X	X	X	X		
E						PLS Regression

5 Conclusions and Future Work

Nowadays, dataset accessibility is vital for Educational Data Mining workflows, it represents the first step, but also a well-documented dataset could enhance the beginning of the EDM investigation and methodology; consequently, investigations can improve and get better results [15].

Based on research, we present the following points to be considered for researchers in EDM field to choose appropriately the datasets to work with:

- Accessibility: keep in mind the source of the dataset to use, need to be a reliable and recognized public repository e.g. UCI ML Repository, Mendeley Data and Kaggle.
- Documentation: having a context or official information about a dataset in advance such as data types, attributes descriptions, and if its public or under a cc licence, allows to improve the beginning of the research.
- Quality: some datasets have missing information and inconsistencies, that requires more complicated preprocessing.
- Up to date: given the speed at which educational data is generated, datasets should be actual with the evaluated population.
- Similarity: dataset may have a likeness with the real data or with the data of the institution that wishes to use it.

The last key factors could influence in aspects like research quality, benchmarking processes, and continuous improvement for future datasets. As a future work, we intend to work in a repository that provides documentation and references for these and future datasets in the higher education domain.

References

1. Adnan, M., Habib, A., Ashraf, J., Mussadiq, S., Raza, A. A., Abid, M., Khan, S. U.: Predicting at-risk students at different percentages of course length for early intervention using machine learning models. Ieee Access, 9, 7519–a7539 (2021)
2. Ampadu, Y. B. (2023). Handling big data in education: A review of educational data mining techniques for specific educational problems. *IntechOpen*. https://doi.org/10.5772/acrt.17
3. Baradwaj, B. K., Pal, S.: Mining educational data to analyze students' performance. arXiv preprint arXiv:1201.3417 (2012)
4. Baker, R. S. J. D. (2010). Data mining for education. *Int. Encycl. Educ., 7*(3), 112–118.
5. Batool, S., Rashid, J., Nisar, M. W., Kim, J., Kwon, H. Y., & Hussain, A. (2023). Educational data mining to predict students' academic performance: A survey study. *Educ. Inf. Technol., 28*(1), 905–971.
6. Breiman, L.: Random forests. Machine learning 45, 5–32 (2001)
7. Cohausz, L. (2022). When probabilities are not enough-A framework for causal explanations of student success models. *J. Educ. Data Min., 14*(3), 52–75.
8. Cortez, P., Silva, A. M. G.: Using data mining to predict secondary school student performance (2008)
9. Delahoz-Dominguez, E., Zuluaga, R., & Fontalvo-Herrera, T. (2020). Dataset of academic performance evolution for engineering students. *Data in brief, 30*, Article 105537.

10. DOF - Diario Oficial de la Federación. (2023, December). Retrieved October 2024, from DOF. https://www.dof.gob.mx/nota_detalle.php?codigo=5712746&fecha=28/12/2023#gsc.tab=0
11. Dutt, A., Ismail, M. A., & Herawan, T. (2017). A systematic review on educational data mining. *Ieee Access, 5*, 15991–16005.
12. Han, J., & Kamber, M. (2006). *Data Mining, Southeast* (Asia). Elsevier Science.
13. Kuzilek, J., Hlosta, M., & Zdrahal, Z. (2017). Open university learning analytics dataset sci. *Data, 4*, Article 170171. https://doi.org/10.1038/sdata.2017.171
14. Lemaitre, M. J. (2017). Quality assurance in Latin America: Current situation and future challenges. *Tuning J. High. Educ., 5*(1), 21–40.
15. Mihaescu, M. C., & Popescu, P. S. (2021). Review on publicly available datasets for educational data mining. *Wiley Interdisciplinary Reviews: Data Mining and Knowledge Discovery, 11*(3), Article e1403.
16. Mohammed, M., Khan, M. B., Bashier, E. B. M.: Machine learning: algorithms and applications. Crc Press (2016)
17. Nurmalitasari, A. L., & Z., & Faizuddin Mohd Noor, M. (2023). Factors influencing dropout students in higher education. *Educ. Res. Int., 2023*(1), 7704142.
18. Rabelo, A., Rodrigues, M. W., Nobre, C., Isotani, S., & Zárate, L. (2024). Educational data mining and learning analytics: A review of educational management in e-learning. *Inf. Discov. Deliv., 52*(2), 149–163.
19. Realinho, V., Machado, J., Baptista, L., & Martins, M. V. (2022). Predicting student dropout and academic success. *Data, 7*(11), 146.
20. Tan, P.N., Steinbach, M., Kumar, V.: Introduction to Data Mining. Pearson: Addison Wesley, Boston (2006)
21. Witten, I. H., & Frank, E. (2005). *Data Mining: Practical Machine Learning Tools and Techniques* (2nd ed.). Elsevier Science.
22. Yılmaz, N., Sekeroglu, B.: Student performance classification using artificial intelligence techniques. In: international conference on theory and application of soft computing, computing with words and perceptions. Cham: Springer International Publishing, pp 596–603 (2019)

Metaheuristics

Application of the Fireworks Algorithm to Design a Fuzzy Controller of an Autonomous Mobile Robot

Juan Barraza, Luis Rodriguez, Fevrier Valdez, Patricia Melin, and Oscar Castillo

Abstract In this paper, we are introducing an application of the Fireworks Algorithm, specifically, we are attacking a problem in the control area. Consequently, in this work, we are using the conventional fireworks algorithm to design a controller using fuzzy logic for a mobile robot system. The Fuzzy Inference System consists in four variables, inputs and outputs, respectively, and nine fuzzy rules. Therefore, the goal of this paper consists on providing the control to the motors based on the design of the fuzzy controller by Fireworks Algorithm. As a result, we tested the performance of the FWA application, minimizing the error compared to a preestablished trajectory of the Autonomous Mobile Robot.

1 Introduction

As we know, intelligent computing [1, 2] has different branches, such as: Artificial Neural Networks (ANNs), and their variants [3, 4], Fuzzy Logic (FL) [5] and Expert Systems (ES) [6, 7], Optimization Algorithms [8] and their variants [9, 10], to mention the most used by the computational scientific community.

In this research work, we focus in two branches: Optimization Algorithms and Fuzzy Logic. Firstly, we show the conventional Fireworks Algorithm (FWA) [11], which is the algorithm that we used for designing a fuzzy controller. Secondly, we applied the FWA algorithm to optimizing a controller, in this case, the algorithm did find the best fuzzy controller for the Robot.

J. Barraza · L. Rodriguez · F. Valdez (✉) · P. Melin · O. Castillo
Tijuana Institute of Technology, TecNM, Tijuana, Mexico
e-mail: fevrier@tectijuana.mx

P. Melin
e-mail: pmelin@tectijuana.mx

O. Castillo
e-mail: ocastillo@tectijuana.mx

O. Castillo and P. Melin (eds.), *Intelligent System Design Based on Soft Computing Models*, Studies in Computational Intelligence 1223,
https://doi.org/10.1007/978-3-031-97309-3_12

Continuing on the branch of Optimization algorithms, we have known a lot of application of them to solve different problems. For example, problems on clustering [12, 13], classification or as this case [14], problems in the control area. Furthermore, we are using the advantages of fuzzy logic to implement a fuzzy controller [15].

On the other hand, the application of fuzzy set theory [16], fuzzy logic [17] and fuzzy reasoning has a considerable weight in a research control area, which is denominated as Fuzzy Logic Control (FLC) [18, 19]. We can find the application of Fuzzy Logic control in diverse areas, such as securities, biomedical instrumentation or any industrial process control. Conventional control techniques compared to Fuzzy Control techniques, it is used for low complexity problems, that is to say, FLC has been best utilized in complex poorly defined control problems.

The robots, specifically designed to move loads autonomously, in a range of industries are named as Autonomous Mobile Robots [20], as we have learned, the robots mentioned are focused on the movement during the trajectory of the robot. Then, the main goal consists on designing a fuzzy controller that improves the movements of a mobile robot system utilizing the Fireworks algorithm.

This paper is organized as follows: Sect. 2 reviews the Conventional FWA- Algorithm their variants and applications [21–31]. In Sect. 3, the Fuzzy Control Design of a Mobile Robot by FWA Algorithm is shown. The experiments and results are shown in Sect. 4; and in Sect. 5 we present the conclusions.

2 FWA Algorithm and Applications

In 2010, a novel swarm optimization was proposed by Y. Tan and Y. Zhu called Fireworks Algorithm [32]. Since they released the algorithm, it has attracted the scientific community to using the algorithm. In this section, we are showing in a general way the steps for executing the Fireworks Algorithm, and, we also show its applications.

2.1 Conventional FWA

FWA is a metaheuristic based on simulating the explosion of the fireworks and, both, fireworks and sparks generated by explosion of fireworks act as possible solutions around of a search space.

After selecting the parameters for algorithm, such as Fireworks number, amplitude and sparks coefficient, dimensions problem, search space, among others. The steps to execute the Fireworks Algorithm (FWA) are the following:

1. **Select *nn* initial locations**: Depending on the problem to be considered, we define a search space to select n n locations.

2. **Set off *NN* Fireworks**: into the search space defined, the algorithm exploit *N N* Fireworks to generate the sparks as possible solutions.
3. **Obtain the locations of sparks**: If the explosion amplitude is big for the firework, the firework generates a low number of sparks; on the contrary, if the explosion amplitude is small for a firework, the firework generates a lot of sparks.
4. **Evaluate the quality of locations**: All possible solutions are evaluated, that it to say, both, fireworks and their sparks are evaluated to select the best solution (location) based on a function evaluation that is implemented according to the problem.
5. **Optimal location found**? If the algorithm found an optimal location then it finishes, on the contrary, the algorithm will be executing until criterion stopping is met.

2.2 First Example of FWA Application.

In this variant of the FWA algorithm, the authors implemented a fuzzy control of the explosion amplitude in a dynamically way. Therefore, the parameter of explosion amplitude stops being constant to switch in a dynamically parameter. The above parameter has a dynamically decrement from 40 to 0. The variant has research work with type-1 fuzzy [23, 25, 26] and type-2 fuzzy [33].

2.3 Second Example of FWA Application.

This variant of FWA was applied to non-negative matrix factorization (NMF).

The scientific research presented and application of the FWA Algorithm, which consists in a new strategy for initialization to the NMF factors, and, they also showed an iterative update strategy in the NMF algorithm [34].

2.4 Third Example of FWA Application

In this variant, the researchers implemented the FWA Algorithm for a classification problem, specifically, an optimization classification model is presented. The authors are called as A self-adaptive fireworks algorithm and denoted as SaFWA, which is responsible to improve the optimization classification model in an efficiently way [35].

2.5 Fourth Example of FWA Application

For this variant, the researchers applied a new FWA method. The FWA is applied in a criteria decision-making problem and the researchers labeled as MCDM-FWA. The purpose of this application of the FWA is solve the selection from alternative locations around the search space. The paper shows a proposed method, a variant of FWA applied as type-1 fuzzy MCDM problems. The interesting part of this research work is the real application problem [36].

3 Fuzzy Control Design of a Mobile Robot by FWA

In this section we explain the application of FWA in fuzzy control. At first, we explained the structure of FWA to solve a fuzzy control problem, and, at second, the model of Fuzzy Control of an Autonomous Mobile Robot (AMR) is offered.

3.1 AFWACP Algorithm

We applied the conventional Fireworks Algorithm to designing a fuzzy controller for the robot system, as we can see, the application of FWA Algorithm is about the control area problems, including, a fuzzy control problem.

We decided to called it as Applied of FWA in Control Problems and we denoted as AFWACP.

The AFWACP method consists of the steps below:

1. **Select *n* initial locations and *N* Fireworks**. For a control problem the search space is variant for each dimension, that is to say, for designing the fuzzy controller the algorithm requires to generates seeds (fireworks and sparks) of 22 dimensions, and, as we mentioned, each dimension has a different range for search space. In Table 1, we are showing the Upper and lower bounds of the 22 dimensions.

2. **Set of *N* at Fireworks at search space**. The fireworks explode based on the amplitude explosion, and consequently, each firework generates a spark as a possible solution. The parameters of coefficient amplitude and sparks, respectively; they are the ones responsible to generates solutions at the search space.

3. **Obtain the locations of sparks**. The following is an example of a firework or spark as possible solution in this fuzzy control problem. Table 2 shows the location of each dimension for this problem.

Table 1 Bounds for each dimension

Dimension	Lower bound	Upper bound
1	−0.3	−0.101
2	−0.03	−0.01
3	−0.1	−0.035
4	0.035	0.1
5	0.01	0.03
6	0.101	0.3
7	−0.3	−0.101
8	−0.03	−0.01
9	−0.1	−0.035
10	0.035	0.1
11	0.01	0.03
12	0.101	0.3
13	−0.75	−0.25
14	−0.1	0.1
15	−0.2	0.2
16	0.25	0.75
17	0.25	0.75
18	−0.75	−0.25
19	−0.1	0.1
20	−0.2	0.2
21	0.25	0.75
22	0.25	0.75

4. **Evaluate the quality of locations**: To evaluate performance of the solutions in the proposed method, for this control problem, we are using a mean square error (MSE); which is shown is Eq. 1:

$$MSE = \frac{1}{n}\sum_{i=1}^{n}(X_i - Y_i)^2 \tag{1}$$

Finally, the AFWACP Algorithm is terminated when the optimal solution is found or the stopping criterion is met.

3.2 *Fuzzy Control of an AMR*

In this section, we show the structure and variables of the autonomous mobile robot model [20].

Table 2 Example of locations (Firework or spark)

Dimension number	Location
1	−0.111492668
2	−0.011054539
3	−0.038427253
4	0.096572747
5	0.028945461
6	0.289507332
7	−0.111492668
8	−0.011054539
9	−0.038427253
10	0.096572747
11	0.028945461
12	0.289507332
13	−0.276363487
14	0.089454605
15	0.17890921
16	0.723636513
17	0.723636513
18	−0.276363487
19	0.089454605
20	0.17890921
21	0.723636513
22	0.723636513

In Fig. 1 the AMR model is shown:

In Eqs. 2 and 3, we are showing the mathematical model to update the movements of the robot:

$$M(q)\dot{v} + C(q, \dot{q})v + Dv = \tau + P(\tau) \tag{2}$$

$$\dot{q} = \begin{bmatrix} cos\theta & 0 \\ sin\theta & 1 \\ 0 & 1 \end{bmatrix} \begin{bmatrix} v \\ w \end{bmatrix} \tag{3}$$

Rodriguez and authors explained a detail the mathematical model in [20].

In Table 3, we outline the name and features of each variable of the FIS for the mobile robot controller:

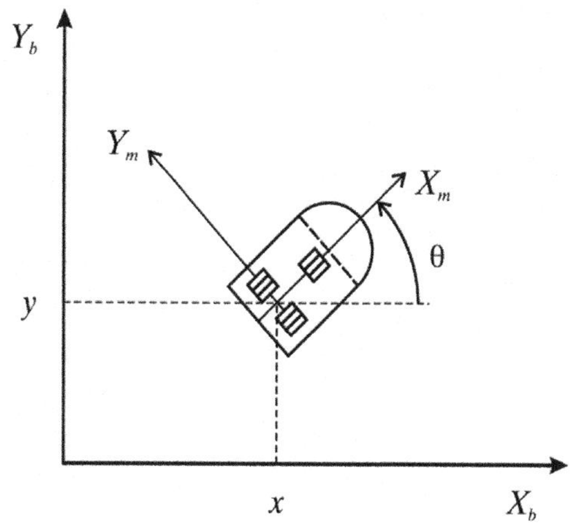

Fig. 1 Graphical representation of the robot

Table 3 Variables of the FIS

Variable	Name	Mean
Input 1	e_V	Error in linear velocity
Input 2	e_W	Error in angular velocity
Output 1	τ_1	Torque 1
Output 2	τ_2	Torque 2

The two input variables consist in three membership functions: two are trapezoidal and one is triangular, which are denoted in the following list:

- Negative (N).
- Zero (Z).
- Positive (P).

On the same way, the two ouput variables are conformed by three triangular memebership functions, which are enlist as:

- Negative (N).
- Zero (Z).
- Positive (P).

In Fig. 2, we are presenting the design fuzzy controller for the AMR.

In the following three figures we show the process to design the fuzzy controller. We start with Fig. 3, which is showing the points of input variables into the FIS for the Fuzzy controller. We can appreciate that the input variables have eleven points, where we have two kind of points, static and dynamic points. The static points are 1, 2 10 and 11. While the dynamic points that the Fireworks Algorithm (AFWACP) is searching are 3’, 5’, 7’ and 9’.

Consequently, in Fig. 4 the points of the output variables are shown. In the same way for the inputs and the output variables there are exists static, dynamic and random

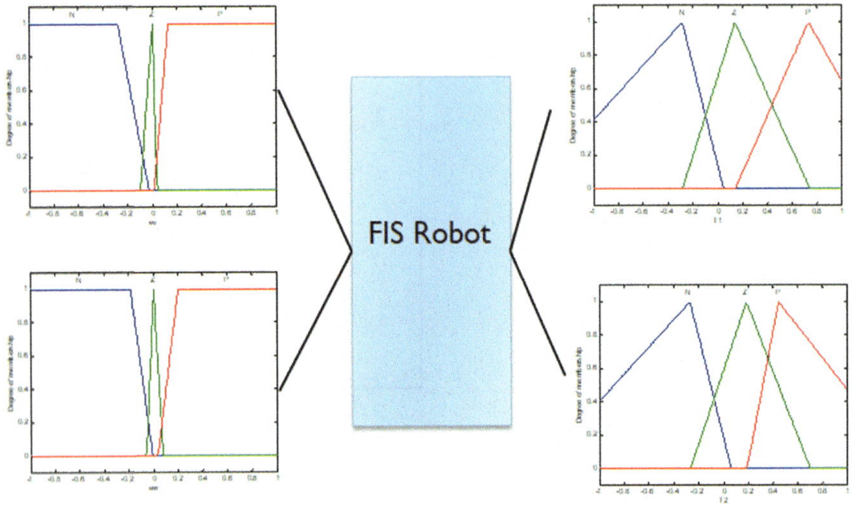

Fig. 2 FIS for the fuzzy controller

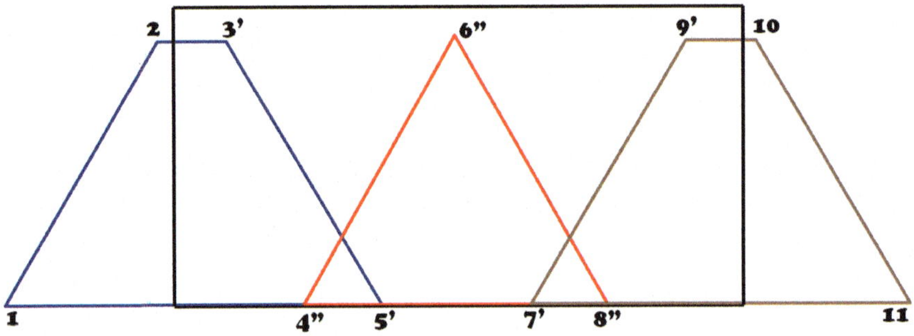

Fig. 3 Points of the inputs variables for the FIS for representation in the fireworks algorithm

points. The static points are 1 and 9; the points 2', 4', 6' and 8' are dynamic points the ones that AFWACP algorithm will search for, and, he 3", 5" and 7"are random points as described above.

In Table 4 we can find the rules that were implemented in the FIS for the fuzzy controller:

Finally, in Fig. 5 we are presenting the ideal trajectory for the robot in this control problem, which has a 'U' form trajectory:

In addition, the performance features of the method will be shown in the following section.

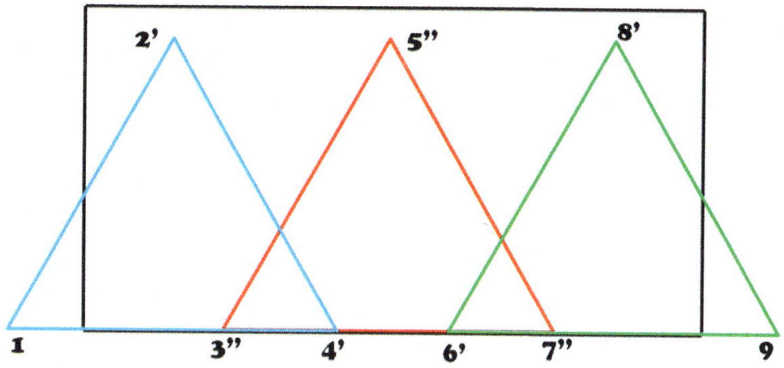

Fig. 4 Points of the outputs variables for the FIS for representation in the fireworks algorithm

Table 4 Fuzzy rules for the FIS

No.	Description
1	If (e_v is Negative) and (e_w is Negative) then (T_1 is Negative) (T_2 is Negative)
2	If (e_v is Negative) and (e_w is Zero) then (T_1 is Negative) (T_2 is Zero)
3	If (e_v is Negative) and (e_w is Positive) then (T_1 is Negative) (T_2 is Positive)
4	If (e_v is Zero) and (e_w is Negative) then (T_1 is Zero) (T_2 is Negative)
5	If (e_v is Zero) and (e_w is Zero) then (T_1 is Zero) (T_2 is Zero)
6	If (e_v is Zero) and (e_w is Positive) then (T_1 is Zero) (T_2 is Positive)
7	If (e_v is Positive) and (e_w is Negative) then (T_1 is Positive) (T_2 is Negative)
8	If (e_v is Positive) and (e_w is Zero) then (T_1 is Positive) (T_2 is Zero)
9	If (e_v is Positive) and (e_w is Positive) then (T_1 is Positive) (T_2 is Positive)

4 Results

In this section, the results by the realized experiments for the fuzzy control problem with proposed method are presented.

It is important to mention that the best result is presented, that it to say, for the control problem we are searching for the best solution with the goal to design a fuzzy controller.

In the following Figures, we are showing the best solution in simulating the trajectory of the AMR.

In Figs. 6 and 7, we show the input variables, and Figs. 8 and 9 shows the output variables, both type of variables belong to the FIS for the fuzzy controller.

The performance measure of the proposed method is given by MSE as we mentioned in above paragraph. For this case we obtained a MSE of 1.4573, which is show in Fig. 10.

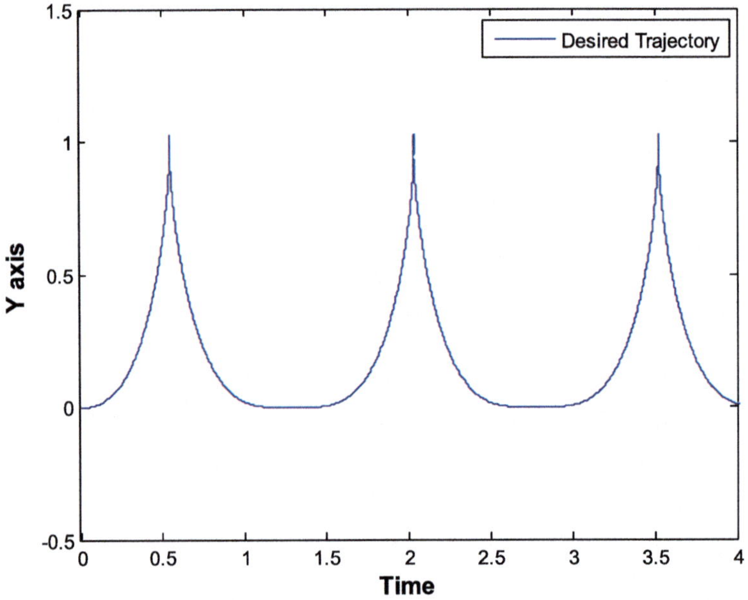

Fig. 5 Simulating desired trajectory for this case study

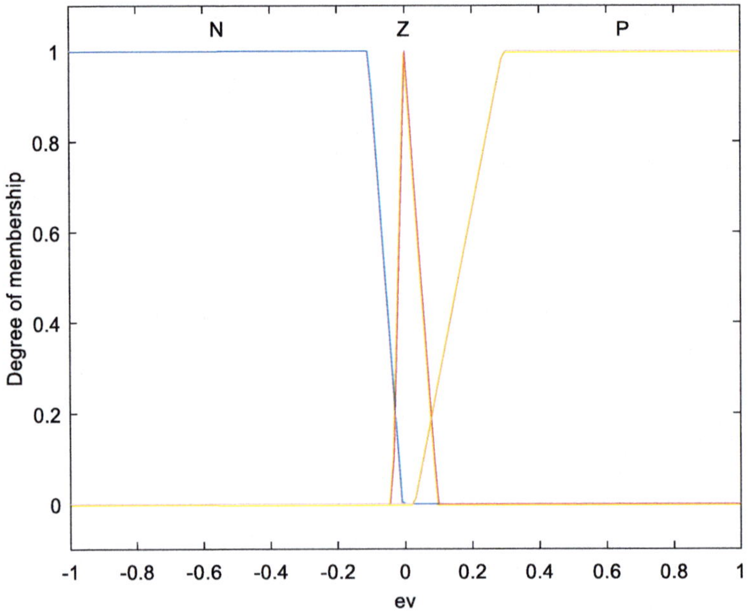

Fig. 6 e_v variable

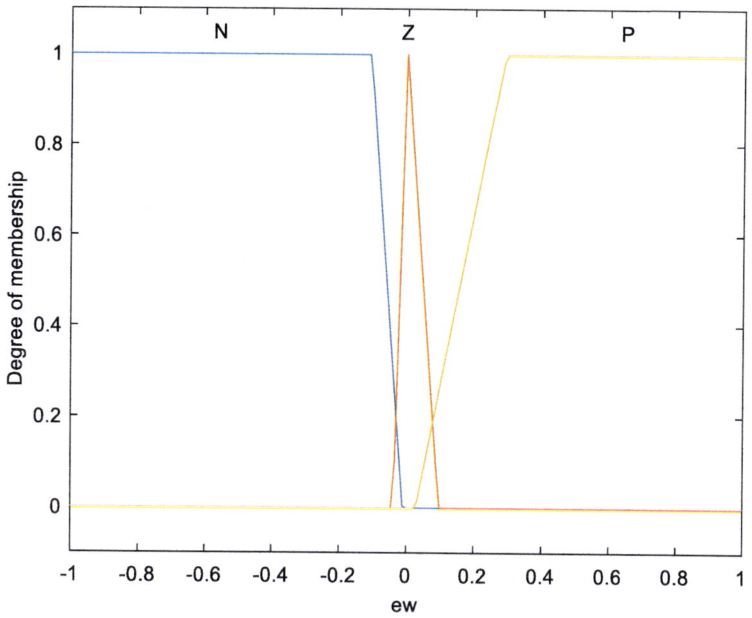

Fig. 7 e_w variable

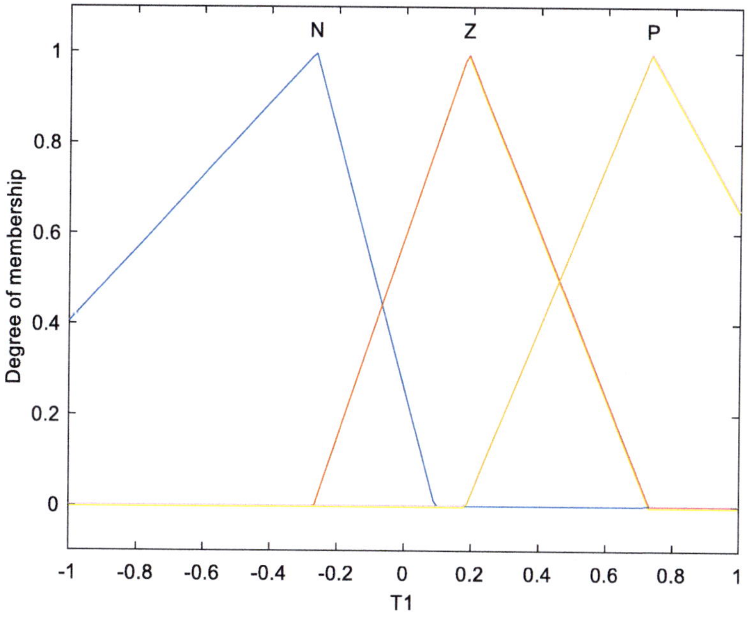

Fig. 8 T_1 variable

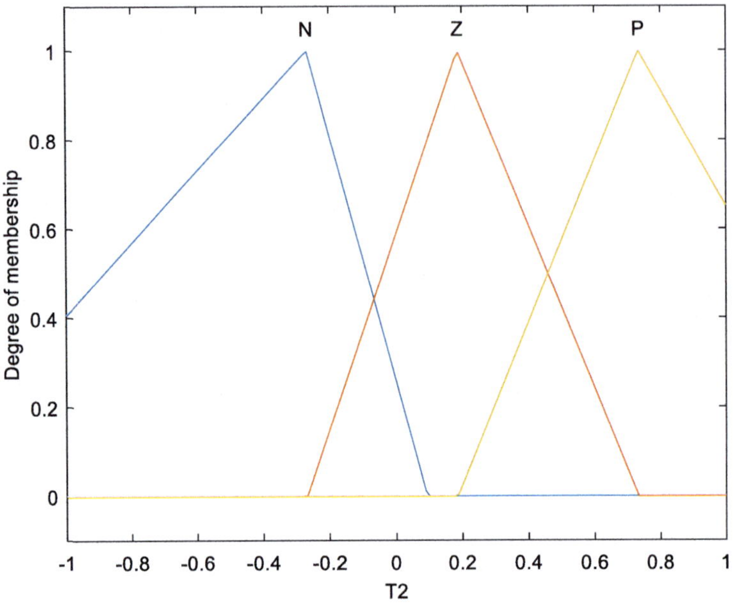

Fig. 9 T_2 variable

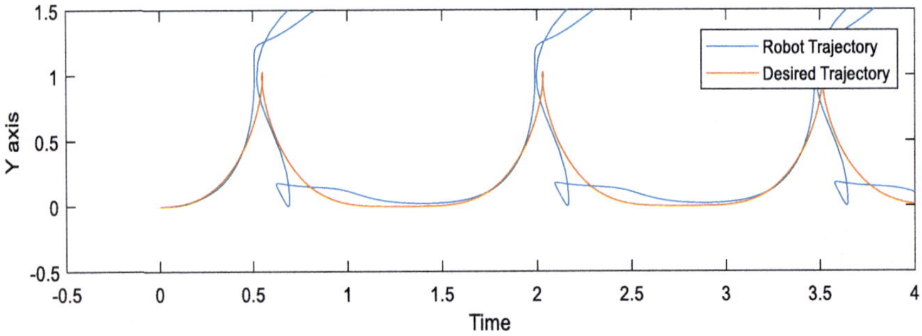

Fig. 10 Best simulating experiment obtained using the fireworks algorithm

As we can appreciate the blue line represents the Robot trajectory of the AMR, simulated with the fuzzy controller given by FWA Algorithm, and the orange line represents of desired trajectory of the Robot.

5 Conclusions

In this recent research work, we have outlined an application of FWA to a Control Problem. The main goal of this paper has been achieved, that it to say, we were able to design, by means of FWA Algorithm a fuzzy controller for providing the control actions to the motors of an Aunotomus Robot Mobile.

Consequently, we could prove that another swarm intelligence technique was adapted to another problem into intelligent computing, in this case for a Fuzzy Control Problem as we mentioned before. Therefore, the main contribution of this work resides mainly in the adaptation of the Optimization Algorithm in a different problem, becasue as we know, FWA Algorithm has advantages that allow it.

The good perfomance of the FWA Algorithm inspires a us to conitnue to apply it in diferent branches of Intelligent Computing.

As future work, we might enhance the performance of FWA in Fuzzy Control problems with dinamically adaptation of parameters using Fuzzy Logic, firstly, with type-1, and secondly, with type-2 Fuzzy Logic. Finally, we also might test FWA in other different problems, such as, Artificial Neural Networks [37], or consider utilizing other metaheuristics [38–40].

References

1. Jukan, A., Masip-Bruin, X., & Amla, N. (2017). Smart computing and sensing technologies for animal welfare: A systematic review. *ACM Computing Surveys (CSUR), 50*(1), 1–27.
2. Vasant, P., Zelinka, I., & Weber, G.W. (Eds.). (2019). *Intelligent computing and optimization* (p. 575). Berlin: Springer International Publishing.
3. Chen, C., Mi, L., Liu, Z., Qiu, B., Zhao, H., & Xu, L. (2021). Predefined-time synchronization of competitive neural networks. *Neural Networks, 142*, 492–499.
4. Soto, J., & Melin, P. (2015). Optimization of the fuzzy integrators in ensembles of ANFIS model for time series prediction: The case of Mackey-glass, *IFSA-EUSFLAT 2015.*
5. Zadeh, L.A. (1989). Knowledge representation in fuzzy logic. *IEEE Transactions on Knowledge and Data Engineering, 1*(1), 89–0084.
6. Rubio, E., Castillo, O. (2017). Interval type-2 fuzzy possibilistic C-means optimization using particle swarm optimization. *Nature-Inspired Design of Hybrid Intelligent Systems* 63–78
7. Bonabeau, E., Dorigo, M., & Theraulaz, G. (1999). Swarm intelligence: from natural to artificial systems: OUP USA
8. Aladwan, F., Alshraideh, M., & Rasol, M. (2015). A genetic algorithm approach for breaking of simplified data encryption standard. *International Journal of Security and Its Applications, 9*(9), 295–304.
9. Melián, B., & Moreno, J. (2003). Metaheuristics: A global vision. *Ibero-American Journal of Artificial Intelligence, 19*, 7–28.
10. Wolpert, D.H., & Macready, W.G. (1997). No free lunch theorems for optimization. *IEEE Transactions on Evolutionary Computation*, pp. 67–82
11. Tan, Y. (2015). *Fireworks Algorithm* (pp. 355–364). Springer-Verlag.
12. Chena, X., Liua, S., Chena, T., Zhang, Z., & Zhang, H. (2012). An improved semi-supervised clustering algorithm for multi-density datasets with fewer constraints. *Procedia Engineering, 29*, 4325–4329.

13. Sanchez, M.A., Castillo, O., Castro, J.R., & Melin, P. (2014). Fuzzy granular gravitational clustering algorithm for multivariate data. *Information Sciences, 279*, 498–511
14. Nerurkar, P., Shirkeb, A., Chandanec, M., & Bhirudd, S. (2018). A novel Heuristic for evolutionary clustering. *Procedia Computer Science, 125*, 780–789.
15. Passino, K.M., Yurkovich, S., & Reinfrank, M. (1998). *Fuzzy control* (vol. 42, pp. 15–21). Reading, MA: Addison-wesley.
16. Zimmermann, H. J. (1996). Fuzzy control. *Fuzzy set theory—and its applications* (pp. 223–264). Springer, Netherlands.
17. Rodríguez, L., Castillo, O., & Soria, J. (2017). A study of parameters of the grey wolf optimizer algorithm for dynamic adaptation with fuzzy logic. *Nature-Inspired Design of Hybrid Intelligent Systems*, pp. 371–390.
18. Siddique, N., & Siddique, N. (2014). *Fuzzy control* (pp. 95–135). Springer International Publishing.
19. Simoes, M., Bose, K., & Spiegel, J. (1997). Fuzzy logic based intelligent control of a variable speed cage machine wind generation system. *IEEE transactions on power electronics, 12*(1), 87–95.
20. Rodriguez, L., et al. (2021). A new metaheuristic optimization algorithm based on a paradigm from physics: String theory, 1657–1675.
21. Abdulmajeed, N.H., & Ayob, M. (2014). A firework algorithm for solving capacitated vehicle routing problem. *International Journal of Advancements in Computing Technology (IJACT), 6*(1), 79–86.
22. Barraza, J., Rodríguez, L., Castillo, O., Melin, P., & Valdez, F. (2018). A new hybridization approach between the fireworks algorithm and grey wolf optimizer algorithm. *Journal of Optimization Research Article, 2018*, 18, Article ID 6495362.
23. Barraza, J., Melin, P., Valdez, F., & González, C. I. (2017). Fuzzy fireworks algorithm based on a sparks dispersion measure. *Algorithms, 10*(3), 83.
24. Barraza, J., Melin, P., Valdez, F., González, C.I., Castillo, O. (2017). Iterative fireworks algorithm with fuzzy coefficients. *FUZZ-IEEE 2017*, Naples, Italy, July 9–12, pp. 1–6
25. Barraza, J., Melin, P., & Valdez, F. (2016). Fuzzy FWA with dynamic adaptation of parameters. *IEEE CEC 2016*, pp. 4053–4060
26. Barraza, J., Valdez, F., Melin, P., & Gonzalez, C. (2017) Fireworks algorithm (FWA) with adaptation of parameters using fuzzy logic. *Nature-Inspired Design of Hybrid Intelligent Systems*, pp. 313–327.
27. Li, J., & S.Z. (2014). Adaptive fireworks algorithm. *IEEE Congress on Evolutionary Computation 2014 (CEC)*, pp. 3214–3221.
28. Liu, J., Zheng, S., & Tan, Y. (2013). The improvement on controlling exploration and exploitation of firework algorithm. In *Advances in Swarm Intelligence*. Springer, pp. 11–23.
29. Tan, Y., & Zheng, S. (2014)) Dynamic search in fireworks algorithm. *Evolutionary Computation (CEC 2014).*
30. Zheng, Y., Song, Q., & Chen, S.-Y. (2013). Multiobjective fireworks optimization for variable-rate fertilization in oil crop production. *Applied Soft Computing, 13*, 4253–4263.
31. Li, J., & Tan, Y. (2019). A comprehensive review of the fireworks algorithm. *ACM Computing Surveys (CSUR), 52*(6), 1–28.
32. Tan, Y., & Zhu, Y. (2010). *Fireworks Algorithm for Optimization* (pp. 355–364). Springer-Verlag.
33. Barraza, J., Valdez, F., Melin, P., & Gonzalez, C.I. (2020). Interval type 2 fuzzy fireworks algorithm for clustering. In Y. Tan (Ed.), *Handbook of Research on Fireworks Algorithms and Swarm Intelligence* (pp. 195–211). IGI Global. https://doi.org/10.4018/978-1-7998-1659-1.ch009.
34. Tan, Y. (2015). FWA application on non-negative matrix factorization. In *Fireworks algorithm: A Novel Swarm Intelligence Optimization Method*, pp. 247–26.
35. Xue, Y., Zhao, B., Ma, T., & Pang, W. (2018). A self-adaptive fireworks algorithm for classification problems. *IEEE Access, 6*, 44406–44416.

36. Mokhtarian, M. N. (2011). A new fuzzy weighted average (FWA) method based on left and right scores: An application for determining a suitable location for a gas oil station. *Computers & mathematics with applications, 61*(10), 3136–3145.
37. Ramírez, E., Melin, P., & Prado-Arechiga, G. (2019). Hybrid model based on neural networks, type-1 and type-2 fuzzy systems for 2-lead cardiac arrhythmia classification. *Expert Systems with Applications, 126*, 295–307.
38. Santiago, B., Dorronsoro, A. J., Nebro, J. J., Durillo, O., & Castillo, H. J. (2019). Fraire, A novel multi-objective evolutionary algorithm with fuzzy logic based adaptive selection of operators: FAME. *Information Sciences, 471*, 233–251.
39. Martínez-Soto, R., Castillo, O., & Aguilar, L. T. (2014). Type-1 and type-2 fuzzy logic controller design using a hybrid PSO–GA optimization method. *Information Sciences, 285*, 35–49.
40. Castillo, O., Melin, P., Ontiveros, E., Peraza, C., Ochoa, P., Valdez, F., & Soria, J. (2019). A high-speed interval type 2 fuzzy system approach for dynamic parameter adaptation in metaheuristics. *Engineering Applications of Artificial Intelligence, 85*, 666–680. https://doi.org/10.1016/j.engappai.2019.07.020

A Review of Unmanned Aerial Vehicles Using PID Control with Genetic Algorithms and Particle Swarm Optimization

Fevrier Valdez

Abstract Unmanned Aerial Vehicles (UAVs) have rapidly emerged as a versatile tool in a variety of fields, including surveillance, mapping, agriculture, and military applications. Central to the efficacy of UAVs systems is their ability to maintain stable and efficient flight control, a task typically handled by control algorithms such as Proportional-Integral-Derivative (PID) controllers. However, UAVs face a broad spectrum of dynamic environments, making conventional PID controllers insufficient to address their nonlinearity and uncertainties. For solving these problems, optimization algorithms, such as Genetic Algorithms (GAs) and Particle Swarm Optimization (PSO), have become popular tools for tuning PID parameters, improving both control precision and adaptability. This paper provides a review of UAVs control systems that incorporate PID controllers tuned using GA and PSO, focusing on their effectiveness, limitations, and future research directions.

Keywords Genetic algorithms · Particle swarm optimization · PID Control · UAVs · Drones · Quadrotor

1 Introduction

Unmanned Aerial Vehicles (UAVs) [1] are widely used in many applications, from military surveillance to environmental monitoring and precision agriculture. The ability of UAVs to perform these tasks is heavily dependent on their control systems. Among the commonly used control strategies, the PID controller is one of the most widely used due to its simplicity and effectiveness in various industrial control systems [2].

However, UAV dynamics are highly non-linear, making traditional PID controllers limited in their ability to maintain stability and accuracy in varying flight conditions. To overcome these challenges, the integration of metaheuristic optimization techniques, such as Genetic Algorithms (GAs) [3–6] and Particle Swarm Optimization

F. Valdez (✉)
Tijuana Institute of Technology/Tecnologico Nacional de Mexico, Tijuana, Mexico
e-mail: fevrier@tectijuana.mx

O. Castillo and P. Melin (eds.), *Intelligent System Design Based on Soft Computing Models*, Studies in Computational Intelligence 1223,
https://doi.org/10.1007/978-3-031-97309-3_13

(PSO) [7–12] have been widely studied. These algorithms help optimize the PID controller parameters to better handle the complex dynamics and uncertainties inherent in UAV operations.

This paper study the use of PID controllers in UAVs, with a specific focus on the application of GAs and PSO as optimization techniques for tuning PID parameters. We also discuss the benefits, drawbacks, and future directions for improving UAVs using these methodologies. This chapter is organized as follows. Section 1, describes the introduction to the chapter. In Sect. 2, the UAV flight control system is described. On the other hand, the optimization algorithms for PID tuning are mentioned in Sect. 3. In Sect. 4, a description of how to build a Vosviewer network is presented. The last Section shows the conclusions and future works. The main contribution of this research work is to show the collected data from the Scopus repository for use with the Vosviewer software, and then to generate several queries and clearly review the most relevant work in the field analyzed in this research.

2 UAV Flight Control Systems

The UAVs Flight Control are an important component of any UAV that manages its stability, navigation, and maneuverability. The system ensures that the UAV can perform its intended tasks, autonomously or under manual control. The following is a detailed description of its components and functions. Flight Controller Unit (FCU): Integrates data from sensors and pilot commands to compute necessary adjustments. Run algorithms to maintain stability, navigation, and mission-specific behavior. Inertial measurement unit (IMU): A sensor package that includes accelerometers, gyroscopes, and sometimes magnetometers [21]. Measures orientation, velocity, and acceleration. Global Navigation Satellite System (GNSS): Typically GPS, GLONASS, or Galileo [22]. Provides precise location and time data. Electronic Speed Controllers (ESCs): Receive signals from the FCU to adjust rotor speeds for direction and altitude control. Communication Module: Facilitates communication between the UAVs and the ground control station (GCS). Uses radio frequencies, Wi-Fi, or satellite links. Power Management System: Monitors and distributes power to various subsystems. Includes battery management to optimize flight duration and safety. Payload Management System: Coordinates sensors, cameras, or other mission-specific equipment. Interfaces with the FCU for seamless integration with flight operations [22].

2.1 UAV Dynamics

UAVs, particularly rotary-wing drones, operate under a set of highly nonlinear and coupled dynamics. These dynamics are influenced by several factors, including wind disturbances, payload variations, and changing environmental conditions. As such, UAV control systems must handle these complexities to ensure stable flight.

The flight dynamics of UAVs is usually modeled through Newton-Euler equations, which govern their translational and rotational motions. These models are crucial for the design and development of control systems capable of ensuring steady and safe UAVs operations under various conditions [23].

2.2 PID Control in UAVs

PID controllers have been the cornerstone of control systems for decades. The three terms–proportional, integral, and derivative–are used to correct for the error between the desired and actual positions or velocities of the UAV.

Proportional control (P): Corrects the error proportionally, responding to the present error. Integral (I) Control: Accumulates the past error over time to eliminate the steady-state error. Derivative (D) Control: Predicts future error by considering the rate of change, thus improving system stability. Despite the simplicity and wide use of PID controllers, tuning the gains (Kp, Ki, Kd) for these controllers can be challenging, especially for UAVs with nonlinear dynamics. This has led researchers to explore optimization methods such as GAs and PSO to enhance the performance of PID controllers in UAV applications [24].

3 Optimization Algorithms for PID Tuning

3.1 Genetic Algorithms

Genetic Algorithms (GAs) are optimization methods inspired by the principles of natural selection and genetics. They work by evolving a population of potential solutions towards an optimal solution. In the context of PID tuning, GAs are employed to find the best combination of proportional, integral, and derivative gains by exploring a wide search space. search space [3, 25]. GAs involve operations such as: Selection: Choosing the best individuals for reproduction. Crossover: Combining the genetic material of two parents to produce offspring with the best traits. Mutation: Introducing random changes to individual allows to explore different parts of the search space.

3.2 Particle Swarm Optimization

PSO was proposed in [7, 8, 10] and the particles represent a possible solution. Each one fly through a search space and the position of each solution is calculated according to its own knowledge and its near particles. The position of a particle i at time step t is calculated by $p_i(t)$. This position can be change adding a velocity $vel_i(t+1)$ to the current position, determinate by Eq. (1):

$$p(t+1) = p_i(t) + vel_i(t+1) \tag{1}$$

In global PSO, the velocity of particle $i$is represented by Eq. (2):

$$vel_{ij}(t+1) = vel_{ij}(t) + c_1 r_{1j}(t)\left[y_{ij}(t) - p_{ij}(t)\right] + c_2 r_{2j}(t)\left[\widehat{y}_j(t) - p_{ij}(t)\right] \tag{2}$$

where $vel_{ij}(t)$ represents the velocity of one individual (particle) i in $j = 1,...,n_x$ at time t, is the position of particle i in j at time t, the personal best position, $y_{ij}(t)$, linked with particle i in dimension j is the best position the particle has visited since the algorithm started, $\widehat{y}_j(t)$is the global best position achieved so far by any particle, c_1 and c_2 are cognitive and social accelerations with static values, and $r_{1j}(t), r_{2j}(t)^{U}$ (0,1) are random values between [0, 1].

4 Vosviewer Network

To create a detailed VosViewer network for the topic of GAs, PSO and PID controllers for UAVs over the last years, we need to focus on keyword clustering and citation relationships across relevant research. This tool allows to extract an overview based on recent studies and trends: Elements of the Network: Visualization: From the Scopus repository, we collect bibliometric data. Keywords like "Genetic Algorithm", "PID Control", and "UAVs" to identify related research. Map co-authorship, co-occurrence of keywords, and citation patterns to form clusters. We can build such a network by importing data into VosViewer and visualizing how keywords and studies interlink, highlighting innovation trends and collaborative networks within this domain. Clusters: We can generate different clusters with the information obtained. Some examples are described below: PID Controller Design for UAVs: This group explores classical PID methods adapted for UAV stability, control, and navigation. Genetic Algorithms (GAs) are frequently used to optimize PID parameters to handle UAVs dynamics and nonlinearities effectively. GA Optimization Techniques: Focuses on the application of GAs to improve the performance of UAV control systems. These include adaptive tuning methods that account for nonlinear and time-varying UAV behaviors. Fixed-Wing vs. Quadcopter UAVs: Differentiates control approaches tailored to specific UAV configurations, where fixed-wing systems and quadcopters have distinct dynamics requiring varied PID tuning methodologies.

4.1 Generated Networks from Scopus Using VosViewer Software

In this section, we present the information collected after making a search on the Scopus repository.

Figure 1, presents the network, strength links, occurrences with the queries used to obtain the information from the Scopus repository. The data were saved in plain text in a csv format. The file was then uploaded to the Vosviewer software to generate the presented result. The setting to represent this information of the queries in this section is as follows for all studied cases: Analysis: Co-authorship. Unit: Authors. Counting method: Full. The minimum number of papers of an author in Fig. 1 was 1. The minimum number of citations was 1. Of the 92 authors, 73 meet the threshold. For each of the 73 authors, the total strength of the coauthor relationships with other authors was calculated. The authors with the highest total link strength were considered. Table 1, shows the top 5 works with the most citations from this study.

Also, we calculate the network with the topic 'UAVs PID Particle Swarm Optimization'. Figure 2, shows the network, strength links, occurrences with the queries used to obtain the information from the Scopus repository. The data were saved in plain text in a csv format. The file was then uploaded to the Vosviewer software

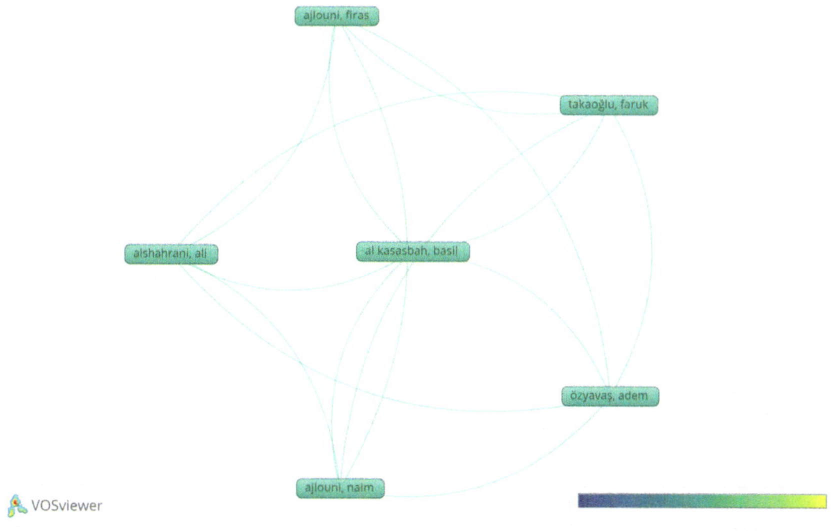

Fig. 1 Illustration for the Topic 'UAVs PID Genetic Algorithms'

Table 1 Review of UAV PID control with genetic algorithms and particle swarm optimization (2014–2024)

Year	Authors	UAV	Objective	Description
2014	Zhang & Wang [13]	Quadrotor	Stabilization of UAVs under disturbances	PID tuning using genetic algorithms
2014	Kumar et al. [14]	Fixed-wing	Optimizing flight path and stability for longer endurance	GA-based PID optimization
2017	González et al. [15]	Quadrotor	Real-time altitude control	Tracking for unmanned aircraft system
2018	Yul et al. [16]	Quadrotor	PSO with PID controller to control a quadrator	PSO with PID controller
2021	Sheta et al. [17]	Quadcopter	Algorithm for the tuning of quadcopter PID parameters	PID controller optimization to stabilize quadcopter movements
2022	Sahrir et al. [18]	Quadcopter	Obstacle avoidance in urban environments	GA-PID integrated with obstacle avoidance algorithms
2023	Muslimov [19]	Quadcopter	Tune the control system of a decentralized UAV	PSO for target encirclement
2024	Rodríguez-Abreo [20]	Multirotor	Optimizer of a fuzzy controller of UAVs	GAs to find the necessary gains to meet the design parameters

to generate the presented result. The minimum number of documents of an author in Fig. 2 was 1. The minimum number of citations was 1. Of the 26 authors, six meet the threshold. For each of the six authors, the total strength of the coauthor relationships with other authors was calculated. The authors with the highest total link strength were considered. Table 3, shows the top 5 works with the most citations from this study. In this case, we can observe less references with respect to Genetic Algorithms. This is because the work is more recent than the study made in Table 2.

5 Conclusions and Future Works

The use of GAs and PSO in the design and optimization of PID controllers for UAVs has demonstrated remarkable potential to improve the overall performance of UAVs in terms of stability, responsiveness, and adaptability. Both GA and PSO have

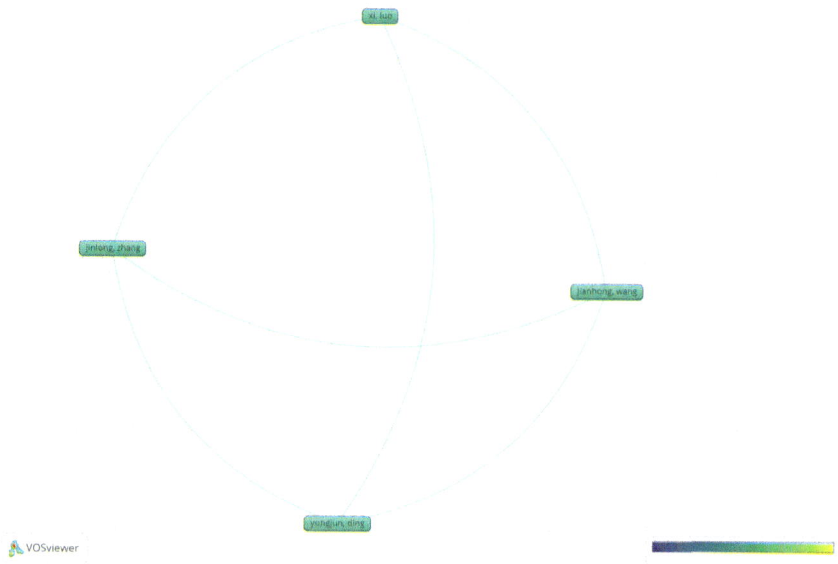

Fig. 2 Illustration for the topic ‘UAVs PID Particle Swarm Optimization’

Table 2 Top 5 of authors with the Topic ‘UAVs PID Genetic Algorithms’

Author	Citations	Reference	Total link strength	Description
Aljahdali et al.	26	[17]	5	PID controller optimization to stabilize quadcopter movements
Bakar et al.	21	[26]	4	Design of solar-powered UAV using GAs
Erkol	18	[27]	0	Attitude controller optimization of UAVs
Passalis et al.	15	[28]	2	Efficient Camera Control for UAVs
Maleki et al.	13	[29]	4	A reliable system design for nondeterministic adaptive controllers in small UAV autopilots

Table 3 Top 5 of authors with the Topic 'UAVs PID Particle Swarm Optimization'

Author	Citations	Reference	Total link strength	Description
Sahrir et al.	11	[30]	1	PSO–PID controller for quadcopter UAV
Liu et al.	0	[31]	3	Reconfiguration motion analysis of an unmanned metamorphic vehicle
Sinan et al.	0	[32]	3	Evaluation of control methods for quadcopter UAVs attitude tracking
Yang et al.	1	[33]	1	Quadrotor attitude control under active disturbance rejection method based on PSO
Yongjun et al.	1	[34]	3	Design of quadcopter attitude controller

proven to be effective in tuning PID parameters, leading to better control performance compared to traditional tuning methods. The optimized controllers show significant improvements in terms of rise time, overshoot, and steady-state error for various UAV dynamics. In addition, these optimization algorithms improve the robustness of the UAV control system against disturbances and parameter uncertainties. The ability of PSO to converge rapidly is particularly advantageous for real-time applications. In addition, it can be effectively scaled to optimize control parameters for UAVs with higher degrees of freedom, such as hexacopters and octocopters, demonstrating their versatility. To further advance the application of GA and PSO in UAV PID controller optimization, many works can be explored for future research and development; for example, hybrid algorithms can be built combining GA and PSO to improve efficiency and adaptability. By addressing these challenges and leveraging advancements in computational intelligence, the application of GA and PSO in UAV PID controller design can continue to evolve, enabling smarter, more efficient, and versatile UAV systems. With this study, it is possible to know the recent research using these two algorithms and with this knowledge is easy to perform experiments and to test the metaheuristics. Also, in future work, a deep study can be performed using other data such as from Web of Science, Google Scholar, etc.

Acknowledgements The author would like to thank CONAHCYT and the Tecnológico Nacional de Mexico/Tijuana Institute of Technology for the support during this research.

References

1. Grlj, C. G., Krznar, N., & Pranjić, M. (2022). A decade of UAV docking stations: A brief overview of mobile and fixed landing platforms. *Drones, 6*(1).
2. Lopez-Sanchez, I., & Moreno-Valenzuela, J. (2023). PID control of quadrotor UAVs: A survey. *Annual Reviews in Control, 56*, Article 100900.
3. Holland, J. H. (1992). Genetic algorithms. *Scientific American*.
4. Whitley, D. (1998). A genetic algorithm tutorial. *Statistics and Computing, 4*, 10.
5. Sequeira, P., Sousa, P. P., Fazendeiro, P. (2011). Genetic algorithms over GPUs: A preliminary assessment. In *International Conf. on Engineering—ICEUBI* (Vol. 1, pp. 1–10).
6. Gaxiola, F., Melin, P., Valdez, F., & Castro, J. R. (2016). Optimization with genetic algorithm and particle swarm optimization of type-2 fuzzy integrator for ensemble neural network in time series. In *2016 IEEE International Conference on Fuzzy Systems (FUZZ-IEEE)* (pp. 2067–2074).
7. Kennedy, J., & Eberhart, R. (1995). Particle swarm optimization. In *Proceedings of ICNN'95—International Conference on Neural Networks* (Vol. 4, pp. 1942–1948).
8. Lorenzo, P. R., Nalepa, J., Kawulok, M., Ramos, L. S., & Pastor, J. R. (2017). Particle swarm optimization for hyper-parameter selection in deep neural networks. In *Proceedings of the Genetic and Evolutionary Computation Conference*, GECCO '17, (pp. 481–488). New York, NY, USA. Association for Computing Machinery.
9. El-Shorbagy, M., & Hassanien, A. E. (2018). Particle swarm optimization from theory to applications. *International Journal of Rough Sets and Data Analysis, 5*, 01.
10. Freitas, D., Lopes, L. G., & Morgado-Dias, F. (2020). Particle swarm optimisation: A historical review up to the current developments. *Entropy, 22*(3).
11. González, B., Melin, P., & Valdez, F. (2020). Particle swarm algorithm for the optimization of modular neural networks in pattern recognition (pp. 59–69). Springer.
12. Shami, T. M., El-Saleh, A. A., Alswaitti, M., Al-Tashi, Q., Summakieh, M. A., & Mirjalili, S. (2022). Particle swarm optimization: A comprehensive survey. *IEEE Access, 10*, 10031–10061.
13. Yang, J., Cai, Z., Lin, Q., Zhang, D., & Wang, Y. (2014). System identification of quadrotor UAV based on genetic algorithm. In *Proceedings of 2014 IEEE Chinese Guidance, Navigation and Control Conference* (pp. 2336–2340).
14. Sharma, R., Rana, K., Kumar, V. (2014). Statistical analysis of GA based PID controller optimization for robotic manipulator. In *2014 International Conference on Issues and Challenges in Intelligent Computing Techniques (ICICT)* (pp. 713–718).
15. González, I., Muñoz Palacios, F., Salazar, S., Espinoza Quesada, E. S., & Lozano, R. (2017). Real-time altitude control for a quadrotor helicopter using a super-twisting controller based on high-order sliding mode observer. *International Journal of Advanced Robotic Systems, 14*, 1729881416687II.
16. Nazaruddin, Y. Y., Andrini, A. D., & Anditio, B. (2018). PSO based PID controller for quadrotor with virtual sensor. *IFAC-PapersOnLine, 51*(4), 358–363. 3rd IFAC Conference on Advances in Proportional-Integral-Derivative Control PID 2018.
17. Sheta, A., Braik, M., Maddi, D. R., Mahdy, A., Aljahdali, S., & Turabieh, H. (2021). Optimization of PID controller to stabilize quadcopter movements using meta-heuristic search algorithms. *Applied Sciences (Switzerland), 11*(14).
18. Sahrir, N. H., & Basri, A. (2022). Modelling and manual tuning PID control of quadcopter (pp. 346–357).
19. Muslimov, T. (2023). Particle swarm optimization for target encirclement by a UAV formation. *Engineering Proceedings, 33*(1).
20. Rodríguez-Abreo, O., Rodríguez-Reséndiz, J., García-Cerezo, A., & García-Martínez, J. R. (2024). Fuzzy logic controller for UAV with gains optimized via genetic algorithm. *Heliyon, 10*(4), Article e26363.
21. Iluk, A. (2023). Flight controller as a low-cost IMU sensor for human motion measurement. *Sensors, 23*(4).

22. Gioia, C. (2023). T-RAIM approaches: Testing with Galileo measurements. *Sensors, 23*(4).
23. Blackwelder, S. (2015). UAV flight dynamics. *International Journal of Aeronautical Science & Aerospace Research,* 81–85.
24. Tehrani, K., & Mpanda, A. (2012). PID control theory.
25. Cabaneros, S. M., & Hughes, B. (2022). Methods used for handling and quantifying model uncertainty of artificial neural network models for air pollution forecasting. *Environmental Modelling & Software, 158*, Article 105529.
26. Bakar, A., Ke, L., Liu, H., Xu, Z., & Wen, D. (2021). Design of low altitude long endurance solar-powered UAV using genetic algorithm. *Aerospace, 8*(8).
27. Erkol, H. O. (2018). Attitude controller optimization of four-rotor unmanned air vehicle. *International Journal of Micro Air Vehicles, 10*(1), 42–49.
28. Passalis, N., Tefas, A., & Pitas, I. (2018). Efficient camera control using 2D visual information for unmanned aerial vehicle-based cinematography.
29. Maleki, K. N., Ashenayi, K., Hook, L. R., Fuller, J. G., & Hutchins, N. (2016). A reliable system design for nondeterministic adaptive controllers in small UAV autopilots.
30. Sahrir, N. H., & Basri, M. A. M. (2023). Pso–pid controller for quadcopter uav: Index performance comparison. *Arabian Journal for Science and Engineering, 48*(11), 15241–15255.
31. Liu, J., Zhao, D., Liu, C., & Xing, J. (2024). Reconfiguration motion analysis and motion quality control of an unmanned metamorphic vehicle. *Control Engineering Practice, 142*.
32. Sinan, S., Bakr, M., Fareh, R., Saad, M., & Bettayeb, M. (2024). Evaluation of control techniques for quadcopter UAV attitude tracking. *Journal of Control and Decision*.
33. Yang, G., Wang, S., Jiang, Y., & Xue, D. (2024). Quadrotor attitude control under active disturbance rejection control method based on particle swarm optimization (pp. 794–798).
34. Yongjun, D., Jianhong, W., Jinlong, Z., & Xi, L. (2024). Design of quadcopter attitude controller based on data-driven model-free adaptive sliding mode control. *International Journal of Dynamics and Control, 12*(5), 1404–1414.

Comparative Analysis of Continuous and Discrete Mycorrhiza Optimization Algorithms

Hector Carreon-Ortiz, Fevrier Valdez, and Oscar Castillo

Abstract Optimization algorithms play a critical role in addressing complex problems across diverse fields such as engineering, economics, and data science. This study conducts an in-depth comparative analysis of the Continuous Mycorrhiza Optimization Algorithm (CMOA) and the Discrete Mycorrhiza Optimization Algorithm (DMOA) by applying them to 36 mathematical benchmark functions. These functions span various types of optimization challenges, including unimodal, multimodal, separable, and non-separable problems, providing a comprehensive assessment of the algorithms' performance. The evaluation criteria include convergence speed, solution accuracy, robustness, and computational efficiency. CMOA, designed for continuous problem spaces, is analyzed for its ability to achieve smooth convergence towards optimal solutions, particularly in problems with continuous search spaces. Its adaptability to parameter variations and the dynamics of the continuous domain is explored in detail. Conversely, DMOA, tailored for discrete optimization tasks, is examined for its effectiveness in managing solution diversity and handling discrete variables with a more precise exploitation of the solution space. The study highlights the distinctive advantages of each algorithm, such as CMOA's superior performance in maintaining stability across complex continuous landscapes and DMOA's flexibility in addressing combinatorial and discrete optimization challenges. Results from the benchmark tests reveal that CMOA generally excels in scenarios with a smooth fitness landscape, showing faster convergence rates and higher precision in continuous domains. On the other hand, DMOA demonstrates greater robustness in discrete scenarios, preventing premature convergence and ensuring diverse solutions. This analysis provides valuable insights into the selection of optimization strategies depending on the problem characteristics, offering a practical guide for researchers in selecting the most suitable algorithm for specific applications. The findings also suggest potential directions for future research, such as hybrid approaches that combine both CMOA and DMOA to tackle increasingly complex optimization problems .

H. Carreon-Ortiz (✉) · F. Valdez · O. Castillo
Tijuana Institute of Technology, TecNM, Tijuana, Mexico
e-mail: hector.carreon17@tectijuana.edu.mx

O. Castillo and P. Melin (eds.), *Intelligent System Design Based on Soft Computing Models*, Studies in Computational Intelligence 1223,
https://doi.org/10.1007/978-3-031-97309-3_14

Keywords Optimization · Continuous · Discrete · CMOA · DMOA

1 Introduction

Optimization is crucial in solving complex problems across a wide range of fields, including engineering design, economic modeling, artificial intelligence, and data science. As the complexity and dimensionality of these problems increase, the demand for effective and efficient optimization algorithms becomes more critical. Nature-inspired optimization algorithms have achieved recognition due to their ability to balance exploration and exploitation. Among these, the Mycorrhiza Optimization Algorithm (MOA), inspired by the relationship between plants and mycorrhizal fungi Fig. 1, has emerged as a promising method for solving optimization problems.

The Continuous Mycorrhiza Optimization Algorithm (CMOA) and its discrete counterpart, the Discrete Mycorrhiza Optimization Algorithm (DMOA), represent adaptations of MOA to different problem domains. CMOA is designed to handle problems with continuous search spaces, making it suitable for tasks where variables can take on a range of values. Its mechanism allows for gradual convergence toward optimal solutions, adapting dynamically to changes in the solution landscape. Conversely, DMOA is tailored for problems with discrete or combinatorial structures,

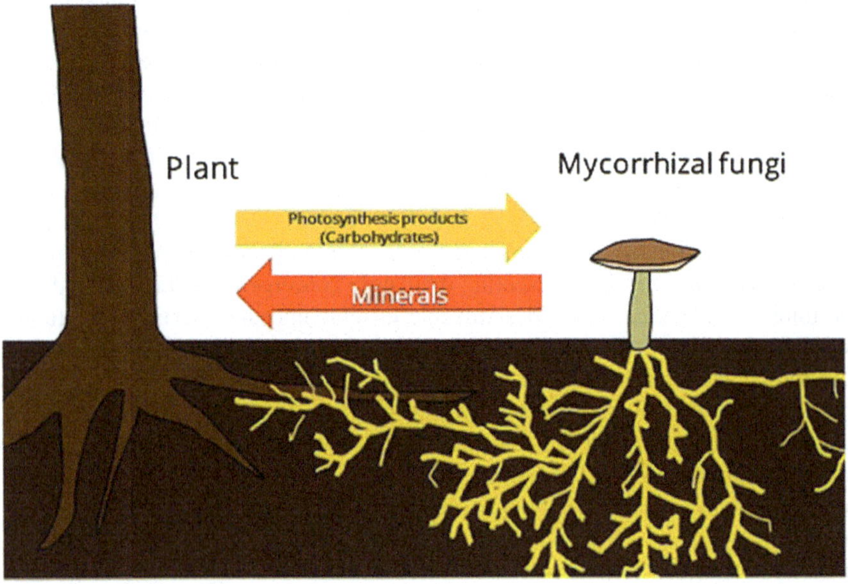

Fig. 1 Mycorrhiza network

where solutions must be selected from a finite set of possibilities. DMOA is particularly effective in maintaining solution diversity, preventing premature convergence, and efficiently navigating through discrete solution spaces.

This study offers a comparative analysis of CMOA and DMOA by evaluating their performance on 36 mathematical benchmark functions. These functions encompass a wide variety of challenges, including multimodal landscapes, separable and non-separable problems, as well as high and low-dimensional spaces. By assessing key metrics, such as convergence speed, accuracy, and efficiency, the study seeks to identify the strengths and limitations of each algorithm in optimization.

The results of this comparison are intended to guide researchers in selecting the appropriate optimization approach for specific applications. Understanding the unique characteristics of CMOA and DMOA enables a more informed choice between these algorithms, depending on the nature of the optimization problem. Furthermore, this study also explores the potential for future advancements, including hybrid approaches that integrate both CMOA and DMOA, aiming to address more complex and diverse optimization challenges.

2 Literature Review

Nature-inspired algorithms has become a major research area in the field of computational intelligence, providing effective solutions to a wide range of optimization problems. These algorithms, such as Genetic Algorithms (GA) [1], Particle Swarm Optimization (PSO) [2], Ant Colony Optimization (ACO) [3], and Differential Evolution (DE) [4], have demonstrated strong capabilities in both continuous and discrete problem domains. However, each algorithm exhibits unique strengths and limitations depending on the nature of the problem, often requiring modifications or hybridization to achieve optimal performance in specific scenarios [5].

The Mycorrhiza Optimization Algorithm (MOA), inspired by the mutualistic interaction between plants and mycorrhizal fungi, Fig. 1 was introduced by Carreon-Ortiz et al. [6] as an alternative approach to existing nature-inspired methods. The primary mechanism of MOA is based on the exchange of resources between plants and fungi, which mimics the sharing of information between solutions in a population. Research has shown that MOA is capable of balancing exploration (searching new areas of the solution space) and exploitation (refining existing solutions), a critical aspect for achieving global optima in complex optimization problems [7]. Studies such as the use of evolutionary and metaheuristic algorithms applied to complex engineering problems with multiple constraints, evaluating their accuracy and efficiency in real large-scale problems [8]. Studies proposing an improved gradient-based optimizer (IGBO) that fine-tunes solutions using inertia and improves the convergence rate in complex optimization problems [9]. Articles such as the "Modified Sine Cosine Algorithm" (MSCA), designed to address complex optimization problems in engineering [10], and another study examining optimization challenges in IoT, big data, and cloud computing contexts [11], analyze the complexity in problems with multiple dimensions and constraints. This work proposes an objective decomposition technique for problems with differentiable and non-differentiable objective functions, facilitating their resolution as multi-objective optimization problems [12], substitution modeling in complex engineering problems, using Kriging models and radial basis functions [13], the review of approximation methods in multi-objective problems, focusing on techniques to improve computational efficiency [14], the exploration of hybrid methods that combine machine learning and optimization to solve complex engineering problems [15], this paper provides an overview of algorithmic techniques for solving complex problems in operational research and decision making [16], study that discusses mathematical models and exact methods applied to optimization problems with multiple variables in engineering and applied sciences [17], research that reviews the use of metaheuristic algorithms in engineering problems, highlighting their advantages for optimization in industrial contexts [18]. Papers using mathematical functions such as: a combination of Harmony Search and Differential Evolution for large-scale optimization, evaluating their performance on 15 CEC 2013 benchmark functions, specifically designed for problems with 1000 continuous variables [19], this article presents 30 mathematical functions for multi-dimensional optimization and compares their complexity with algorithms such as the Genetic Algorithm [1], Particle Swarm Optimization [2], and Sequential Quadratic Programming (SQP) [20], which allows identifying the suitability of each algorithm for specific optimization problems [21].

While MOA is effective in various continuous optimization tasks, its direct application to discrete or combinatorial problems presents challenges. This has led to the development of the DMOA, a variant specifically designed to handle problems where variables are limited to a finite set of values. DMOA adapts the principles of MOA to discrete search spaces by incorporating mechanisms that ensure precise adjustments in the population, preventing premature convergence and maintaining diversity [22–24]. Comparative studies that have been done with DMOA and other studies [25–27] show that DMOA has been better in certain combinatorial problems, demonstrating its effectiveness in maintaining its balance between local and global search capabilities.

The CMOA retains the original structure of MOA but focuses on enhancing convergence speed and precision in continuous search spaces. In research studies [28], they have demonstrated their competitive advantage in terms of convergence efficiency compared to classical algorithms such as FPA [29], BA [30], CS [31], GA [31, 33], HS [31], PSO [31, 33], DE [32], SPSO [33] and WSA [33]. The ability of CMOA to dynamically adapt parameters during the optimization process has been emphasized as a key factor in its successful application to real-world continuous optimization problems.

The present study aims to provide an understanding of how these algorithms perform under various optimization challenges, offering insights into their practical applicability and potential for hybridization.

3 Methodology

The primary goal of this research is to conduct a detailed analysis between the CMOA and the DMOA, applying them to a comprehensive set of 36 benchmark functions. The methodology followed in this study is structured in the following stages:

3.1 Selection of Test Functions

A set of 36 benchmark functions widely utilized in the optimization community was chosen to evaluate the metaheuristics. These functions encompass diverse characteristics:

Unimodal Functions: To test the exploitation ability and precision in finding the global optimum.

Multimodal Functions: To evaluate the ability to avoid local optima and maintain diversity in the population.

Separable and Non-Separable Functions: To analyze how variable interactions are handled.

High-Dimensional Functions: To test scalability and efficiency in larger search spaces.

3.2 Implementation of the Algorithms

CMOA Algorithm

- Implemented following the original MOA model adapted for continuous search spaces.
- Includes operations such as symbiosis, competition, and colonization applied to continuous variables.
- Parameters were established, such as population size, number of iterations, nutrient exchange rates, among others.

DMOA Algorithm

- The MOA was adapted to work in discrete spaces, modifying operations to handle discrete or integer variables.
- Specific operators were adjusted to maintain the validity of solutions in discrete domains.
- Similar parameters to the CMOA were defined but adjusted to the discrete nature of DMO.

3.3 Experimental Setup

Common Parameters

- Both algorithms used the same number of iterations and population size to ensure a fair comparison.
- Preliminary tests were conducted to determine optimal values for each algorithm's specific parameters.
- Multiple Runs:
- Each algorithm was executed 30 times per function to obtain statistically significant results.
- Random seeds were controlled to ensure the reproducibility of the results.

3.4 Evaluation Procedure

Initialization

- An initial random population was generated within the defined limits for each function.
- For the CMOA, variables were initialized as continuous values; for the DMOA, as allowed discrete values.

Algorithm Iteration

- In each iteration, the specific operations of each algorithm (symbiosis, nutrient exchange, competition, etc.) were applied.
- Solutions were evaluated using the corresponding objective function.
- Stopping Criterion:
- Algorithms terminated upon reaching the maximum number of iterations or if the change in the optimal value was less than the predefined tolerance.

3.5 Data Collection and Analysis

Performance Metrics

- Recorded the best value found, mean value, standard deviation, and execution time.
- Convergence curves were analyzed to visualize the algorithms' behavior over time.
- Statistical Analysis:
- Statistical tests (such as Student's t-test and ANOVA) were undertaken to estimate the significance.
- The robustness and stability of the algorithms were evaluated through the variability of results across multiple runs.

3.6 *Comparison of Results*

Overall Performance

- Results obtained by the CMOA and DMOA were compared across all functions, identifying general trends.
- Algorithms were ranked based on their average performance and consistency.
- Analysis by Function Type:
- Performance was evaluated in groups of functions with similar characteristics (e.g., multimodal, high-dimensional).
- Strengths and weaknesses of each algorithm were identified according to the problem type.
- Comparison with Previous Studies:
- Results were contrasted with those reported for similar algorithms.
- Highlighted the study's contribution in relation to previous research.

3.7 *Tools and Development Environment*

- Algorithms were programmed in a language suitable for numerical calculations and optimization (e.g., MATLAB or Python).
- Specialized libraries and functions were used to ensure efficiency and precision of operations.

This rigorous methodology enabled a comprehensive evaluation of the CMOA and DMOA algorithms, providing a solid foundation to compare their performance and draw meaningful conclusions. By addressing different types of functions and controlled experimental conditions, the study offers key insights into the effectiveness of each algorithm in various optimization scenarios.

The Continuous Mycorrhiza Optimization Algorithm (CMOA) is an optimization technique based on the continuous interaction between plants and mycorrhiza networks (MN) that simulate natural processes of symbiosis and competition. Inspired by the Lotka-Volterra equations, this algorithm employs biological models to improve the adaptation and evolution of solutions throughout the iterative process.

The flowchart of CMOA Fig. 2 begins with the definition of parameters, where dimensions, number of iterations, and other relevant parameters are established. Next, the population of plants and mycorrhiza networks is initialized, followed by the evaluation of the plants' fitness to identify the best solutions in the population.

The structure of the algorithm relies on a series of biological operators that include three fundamental models: the defense model (predator–prey) based on the Lotka-Volterra equations, the resource exchange (cooperative model), and colonization (competitive model). These operators simulate ecological dynamics that facilitate the exploration and exploitation.

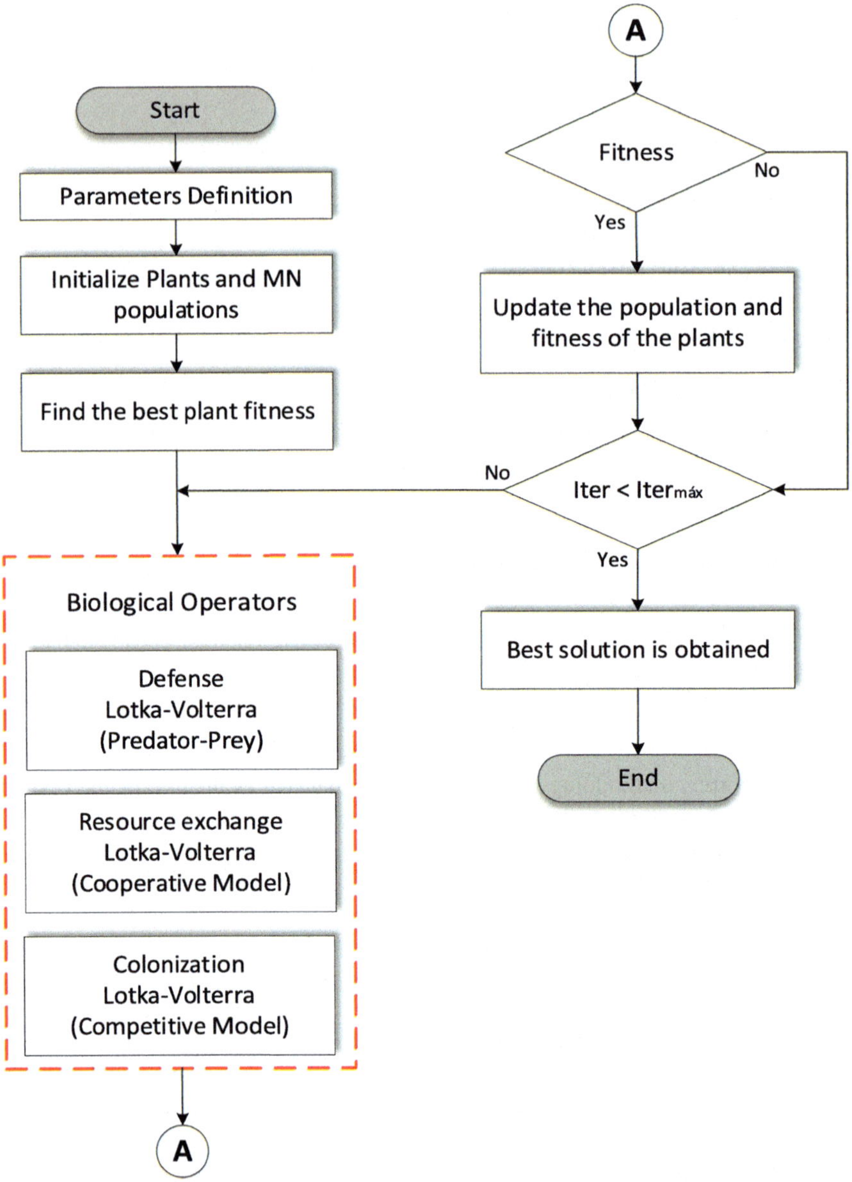

Fig. 2 CMOA flowchart

The process continues iteratively until the maximum number of iterations is reached. During each iteration, the solutions are evaluated, and both the population and the fitness of the plants are updated. Once the iterations are completed, the algorithm converges to the best solution obtained for the optimization problem.

DMOA is a method inspired by the symbiosis and competition between plants and mycorrhizal fungi. This algorithm simulates ecological interactions that enable exploration and exploitation to find optimal solutions.

The flowchart in Fig. 3 begins with parameter initialization, including problem dimensions, number of epochs iterations, and other relevant settings. Next, the population of plants and mycorrhizae is initialized, and the initial best fitness values are determined for both populations.

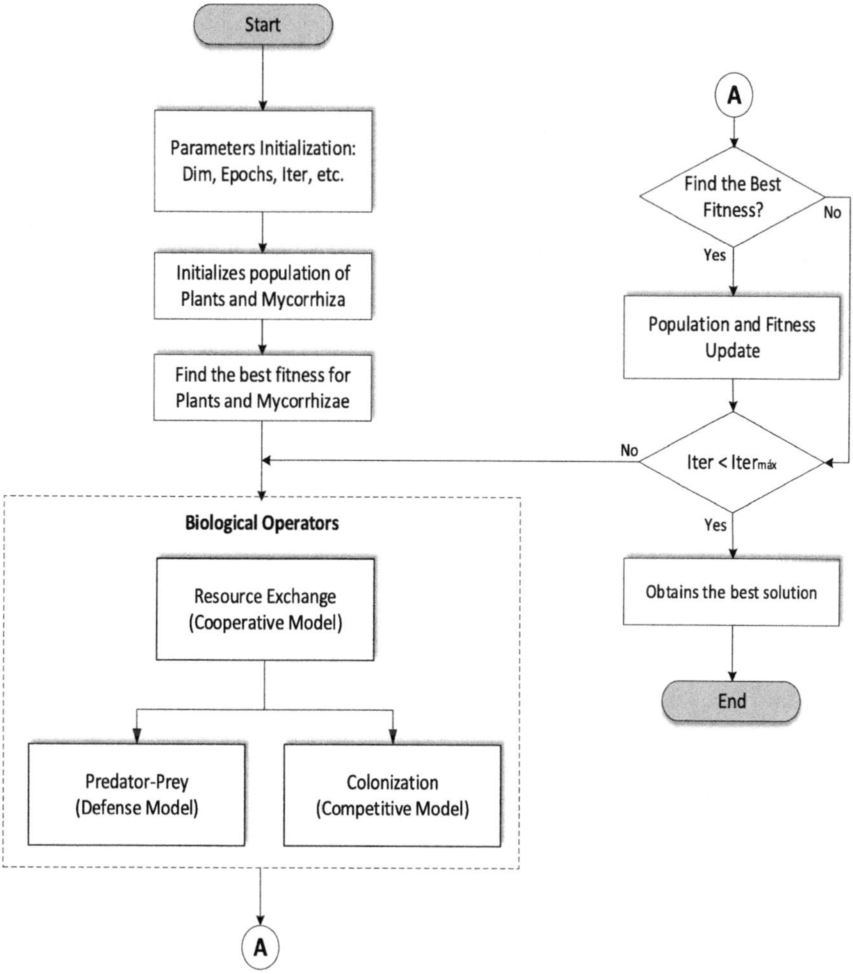

Fig. 3 DMOA flowchart

The biological operators section represents the core of the algorithm and consists of the resource exchange model (cooperative model), which facilitates collaboration between plants and mycorrhizae, and two additional models: predator–prey (defense model) and colonization (competitive model), which add defensive and competitive dynamics to the interactions.

The iterative process of the algorithm continues until a termination criterion is met or obtaining a satisfactory solution. In each iteration, the algorithm evaluates and updates the plant and mycorrhizae populations based on the best fitness value found. Finally, the process concludes when the best solution is obtained or the termination criteria are met.

4 Implementation

CMOA is inspired by the mutualistic relationship between plants and mycorrhizal fungi. In this algorithm, the elements are defined as follows.

- Plants: Represent possible solutions to the optimization problem.
- Fungi: Assist plants in absorbing nutrients, thereby improving the solutions.

The algorithm alternates between exploration and exploitation phases to find the global optimum.

DMOA adapts the principles of Continuous Mycorrhiza Optimization Algorithm to discrete search spaces. Solutions are represented as discrete combinations, and updates are made with operators suited for discrete variables.

Both algorithms follow similar principles but are tailored to different types of problems:

- CMOA is suitable for continuous optimization problems where variables can take any value within a range.
- DMOA is designed for discrete or combinatorial problems where variables can only assume specific values (e.g., 0 or 1).

The implementing CMOA and DMOA allows us to explore how algorithms inspired by natural processes can be adapted to different types of optimization problems. By understanding and comparing these algorithms, we can identify opportunities to enhance their performance and applicability in various scenarios.

Table 1 shows a set of 36 benchmark functions used in the CEC2013 competition, which are commonly employed to assess the performance of optimization algorithms. Each function is characterized by its name, search range, and nature. The following details are provided:

- Function: Includes names of well-known functions such as Sphere, Rosenbrock, Griewank, Ackley, which are classic in the evaluation of optimization problems.

Table 1 36 Benchmark functions

NF	Function	Range	Nature
F1	Sphere	[–5.12, 5.12]	U
F2	Rosenbrok	[–5, 10]	U
F3	Griewank	[–600, 600]	M
F4	Rastrigin	[–5.12, 5.12]	M
F5	Ackley	[–32.768, 32.768]	M
F6	Dixon-price	[–10, 10]	U
F7	Michalewicz	[0, π]	M
F8	Powell	[–4, 5]	U
F9	RHE - Rotate Hyper Ellipsoid	[–65.536, 65.536]	U
F10	Schwefel	[–500, 500]	M
F11	Styblinski-Tang	[–5, 5]	U
F12	SDP - Sum Different Powers	[–1, 1]	M
F13	Sum squares	[–10, 10]	U
F14	Trid	[$-d^2$, d^2]	U
F15	Zakharov	[–5, 10]	U
F16	Bukin No 6	[–15, –5]	U
F17	Cross-in Tray	[–10, 10]	M
F18	Drop-Wave	[–5.12. 5.12]	M
F19	Eggholder	[–5.12, 5.12]	M
F20	Beale	[–4.5, 4.5]	U
F21	Holder-Table	[–10, 10]	M
F22	Branin	[–5, 10]	M
F23	Levy	[–10,10]	M
F24	Levy 13	[–10, 10]	M
F25	Schaffer 2	[–100, 100]	M
F26	Schaffer 4	[–100, 100]	M
F27	Shubert	[–10, 10]	M
F28	Bohachevsky 1	[–100, 100]	M
F29	Bohachevsky 2	[–100, 100]	M
F30	Bohachevsky 3	[–100, 100]	M
F31	Booth	[–10, 10]	U
F32	Matyas	[–10, 10]	U
F33	Mccormick	[–1.5, 4]	U
F34	Easom	[–100, 100]	U
F35	Goldstein-Price	[–2, 2]	M
F36	Three-Hump Camel	[–5, 5]	M

- Range: Specifies the search interval for each function. For instance, the Sphere function has a range of [−5.12, 5.12], while the Griewank function features a broader range of [−600,600].
- Nature: Indicates whether the function is unimodal (U) or multimodal (M). Unimodal functions (U) have a single global optimum, whereas multimodal functions (M) contain multiple local optima, increasing the complexity of the problem.

Table 1 is crucial for comparing and analyzing the performance of optimization algorithms, as it presents a variety of challenges in terms of search surfaces and function complexity.

5 Results

This section presents and analyzes the results of applying CMOA and DMOA to 36 benchmark mathematical functions for 30 dimensions.

These benchmark functions were selected to evaluate the capabilities of both algorithms across different types of optimization scenarios, including unimodal, multimodal, separable, and non-separable problems. The diversity of these functions enables a comprehensive evaluation of each algorithm's performance in terms of convergence speed, accuracy, robustness, and computational efficiency.

Each benchmark function was evaluated through multiple runs to obtain average values and standard deviations, providing a statistical view of each algorithm's consistency and effectiveness across different configurations. Additionally, hypothesis testing was performed to assess the statistical significance of the observed performance differences between CMOA and DMOA, enabling an objective comparison of their optimization capabilities in continuous and discrete search spaces. Through this comparative analysis, the aim was to identify the strengths and limitations of CMOA and DMOA based on the specific challenges presented by different types of optimization problems. These results not only provide a clear perspective on the applicability of each algorithm but also suggest potential areas for improvement and adaptation in the development of optimization techniques in future optimization studies.

Table 2 presents the results of a comparative analysis between the Continuous Mycorrhiza Optimization Algorithm and DMOA across 36 mathematical functions (f1 to f36) in a 30-dimensional space. For each function, the table includes the mean (Mean1 for CMOA and Mean2 for DMOA) and standard deviation (SD1 for CMOA and SD2 for DMOA) values, as well as the results of hypothesis testing.

The hypothesis testing section provides the Z-value, p-value, and the significance evaluation (E), where "N" indicates that CMOA achieved statistically better results, "Y" suggests that DMOA performed better, and "—" denotes no statistically significant difference between the two algorithms. Out of the 36 functions tested, CMOA demonstrated superior performance in 33 functions (indicated by

Table 2 Hypothesis testing for comparative performance of CMOA and DMOA across 36 benchmark functions in 30-dimensional space

30 D	CMOA		DMOA		Hypothesis testing		
F	Mean1	SD1	Mean2	SD2	Z	p	E
f1	8.11×10^{-9}	4.79×10^{-9}	1.87×10^{-8}	1.35×10^{-8}	4.057	< 0.0001	N
f2	1.64×10^{-9}	1.39×10^{-9}	2.69×10^{-9}	2.53×10^{-9}	1.988	0.023	N
f3	1.44×10^{-9}	1.21×10^{-9}	1.00×10^{-9}	8.92×10^{-10}	−1.606	0.054	Y
f4	8.13×10^{-9}	4.42×10^{-9}	1.61×10^{-8}	1.46×10^{-8}	2.856	0.002	N
f5	6.58×10^{-9}	3.32×10^{-9}	2.02×10^{-8}	1.78×10^{-8}	4.113	< 0.0001	N
f6	6.76×10^{-9}	4.01×10^{-9}	1.82×10^{-8}	1.88×10^{-8}	3.245	0.001	N
f7	5.70×10^{-9}	3.18×10^{-9}	1.95×10^{-8}	1.53×10^{-8}	4.859	< 0.0001	N
f8	9.13×10^{-9}	2.87×10^{-9}	1.89×10^{-8}	1.41×10^{-8}	3.705	0.000	N
f9	6.95×10^{-9}	3.99×10^{-9}	1.88×10^{-8}	1.66×10^{-8}	3.809	< 0.0001	N
f10	8.42×10^{-9}	4.41×10^{-9}	2.04×10^{-8}	1.90×10^{-8}	3.353	0.000	N
f11	8.11×10^{-9}	4.14×10^{-9}	1.61×10^{-8}	1.51×10^{-8}	2.806	0.003	N
f12	6.34×10^{-9}	3.72×10^{-9}	1.65×10^{-8}	1.62×10^{-8}	3.332	0.000	N
f13	8.02×10^{-9}	4.68×10^{-9}	1.68×10^{-8}	1.31×10^{-8}	3.470	0.000	N
f14	7.02×10^{-9}	3.78×10^{-9}	2.14×10^{-8}	1.68×10^{-8}	4.561	< 0.0001	N
f15	9.58×10^{-9}	4.28×10^{-9}	1.84×10^{-8}	1.62×10^{-8}	2.879	0.002	N
f16	8.75×10^{-9}	4.63×10^{-9}	1.87×10^{-8}	1.47×10^{-8}	3.555	0.000	N
f17	1.73×10^{-9}	1.24×10^{-9}	2.16×10^{-9}	1.89×10^{-9}	1.026	0.152	N
f18	9.52×10^{-11}	1.13×10^{-10}	2.40×10^{-11}	2.89×10^{-11}	−3.330	< 0.0001	Y
f19	9.63×10^{-9}	4.01×10^{-9}	1.62×10^{-8}	1.39×10^{-8}	2.499	0.006	N
f20	8.04×10^{-9}	4.75×10^{-9}	1.53×10^{-8}	1.36×10^{-8}	2.753	0.003	N
f21	2.95×10^{-9}	1.93×10^{-9}	4.21×10^{-9}	4.4567E-09	1.416	0.078	N
f22	8.31×10^{-9}	4.30×10^{-9}	1.6442E-08	1.51×10^{-8}	2.843	0.002	N
f23	1.00×10^{-8}	4.81×10^{-9}	1.78×10^{-8}	1.3819E-08	2.905	0.002	N
f24	9.08×10^{-9}	4.24×10^{-9}	2.1665E-08	1.51×10^{-8}	4.406	< 0.0001	N
f25	1.77×10^{-10}	1.86×10^{-10}	3.98×10^{-10}	3.72×10^{-10}	2.911	0.002	N
f26	2.93×10^{-7}	2.68×10^{-8}	0.00010794	2.581E-09	21,884.866	< 0.0001	N
f27	8.15×10^{-9}	4.08×10^{-9}	1.9185E-08	1.66×10^{-8}	3.533	0.000	N
f28	7.36×10^{-9}	4.10×10^{-9}	2.4776E-08	1.95×10^{-8}	4.788	< 0.0001	N
f29	8.07×10^{-9}	4.05×10^{-9}	1.9361E-08	1.5029E-08	3.972	< 0.0001	N
f30	8.3477E-09	4.715E-09	1.6384E-08	1.2825E-08	3.221	0.001	N
f31	8.0987E-09	4.17×10^{-9}	1.4439E-08	1.51×10^{-8}	2.211	0.014	N
f32	1.06×10^{-8}	5.4363E-09	1.1822E-08	1.2152E-08	0.483	0.315	N

(continued)

Table 2 (continued)

30 D	CMOA		DMOA		Hypothesis testing		
F	Mean1	SD1	Mean2	SD2	Z	p	E
f33	6.7793E-09	4.0712E-09	1.5511E-08	1.65×10^{-8}	2.808	0.002	N
f34	6.3518E-27	1.0753E-26	5.1931E-29	3.5072E-29	–3.209	< 0.0001	Y
f35	8.4829E-09	4.75×10^{-9}	1.7279E-08	1.5478E-08	2.976	0.001	N
f36	7.2682E-09	4.03×10^{-9}	1.62×10^{-8}	1.5447E-08	3.068	0.001	N

'N'), while DMOA showed better results in only 2 functions (indicated by 'Y'), with one function showing no significant difference. The p-values highlight the statistical significance of these comparisons, with many values being extremely low ($p < 0.0001$), underscoring the robustness of the observed differences.

This analysis reveals a clear performance advantage of CMOA over DMOA across most benchmark functions in a 30-dimensional space, suggesting that CMOA is generally more effective in optimizing these functions within this experimental setup.

6 Conclusion

The experiments conducted demonstrate that CMOA consistently outperforms DMOA in terms of accuracy and stability across most benchmark functions and dimensions tested. Table 2, which presents the hypothesis testing results, reinforces these findings by showing that CMOA significantly outperformed DMOA in 33 out of the 36 functions tested in the 30-dimensional space. The extremely low p-values ($p < 0.0001$ in many cases) confirm the statistical significance of the observed differences, supporting the superiority of CMOA with a high level of confidence. DMOA outperformed CMOA in only two functions, while one function showed no statistically significant difference between the two algorithms.

These results lead us to conclude that CMOA is the more suitable algorithm for continuous optimization problems in 30-dimensional space, especially in applications where precision and consistency in results are essential. CMOA's ability to maintain low error and variability values indicates that it is more stable and effective in continuous optimization contexts. On the other hand, DMOA's relatively lower performance suggests it may not be the best choice for high-dimensional continuous problems, although it could be useful in discrete or combinatorial optimization problems.

CMOA demonstrates a remarkable ability to converge towards near-optimal solutions faster than DMOA, as reflected in consistently lower mean values across most of the functions evaluated. In comparison, DMOA tends to converge more slowly towards optimal solutions, as indicated by its higher mean values in almost all evaluated functions.

CMOA demonstrates a clear advantage over DMOA in the continuous optimization of mathematical functions across multiple dimensions, which is a relevant finding for researchers and practitioners seeking efficient and reliable optimization algorithms in high-dimensional contexts. These results also suggest potential benefits in exploring hybrid variations that combine the strengths of both algorithms to address a broader range of problems. In future we could try applications [34] or compare with other algorithms [35–37].

References

1. Holland, J. (1975). *Adaptation in Natural and Artificial Systems*. University of Michigan Press.
2. Kennedy, J., Eberhart, R.: Particle swarm optimization. In: Proceedings of the IEEE international conference on neural networks, Vol 4. IEEE, pp 1942–1948) (1995)
3. Dorigo, M., & Stützle, T. (2004). *Ant Colony Optimization*. The MIT Press.
4. Storn, R., & Price, K. (1997). Differential evolution – a simple and efficient heuristic for global optimization over continuous spaces. *J. Glob. Optim., 11*(4), 341–359.
5. Yang, X.S. (Ed.) Recent advances in swarm intelligence and evolutionary computation Vol 585. Springer International Publishing (2015). https://doi.org/10.1007/978-3-319-13826-8.
6. Carreon-Ortiz, H., Valdez, F., & Oscar, C. (2023). Mycorrhiza optimization algorithm: A new metaheuristic for global optimization. *Applied Soft Computing, 123*, Article 108897.
7. Xu, X.: Enhancing optimization through innovation: The multi-strategy improved black widow optimization algorithm (MSBWOA). arXiv preprint arXiv:2312.13395 (2023)
8. Premkumar, M., Johnson, C.: Algorithms for real-world complex engineering optimization problems. Algorithms (2022)
9. Moin, L., & Alsisi, R. H. (2023). An improved gradient-based optimization algorithm for solving complex optimization problems. *Processes, 11*(2), 498.
10. Shang, C., Zhou, T., & Liu, S. (2022). Optimization of complex engineering problems using modified sine cosine algorithm. *Scientific Reports, 12*, 20528.
11. Zhan, Z.-H., Shi, L., Tan, K. C., & Zhang, J. (2022). A survey on evolutionary computation for complex continuous optimization. *Artif. Intell. Rev., 55*(1), 59–110. https://doi.org/10.1007/s10462-021-10042-y
12. Liu, J., Pan, Y.: On solving complex optimization problems with objective decomposition. IEEE Xplore (2023)
13. Zhang, X., Wang, J.: Surrogate-based optimization for complex engineering problems. IEEE Transactions on Systems, Man, and Cybernetics (2022)
14. He, S., Wu, H.: Approximation methods for multiobjective optimization problems: A survey. INFORMS J. Comput (2022)
15. Azevedo, B. F., Rocha, A. M. A. C., & Pereira, A. I. (2024). Hybrid approaches to optimization and machine learning methods: A systematic literature review. *Machine Learning, 113*(7), 4055–4097. https://doi.org/10.1007/s10994-023-06467-x
16. Silva, C., Ribeiro, R., Gomes, P.: Algorithmic optimization techniques for operations research problems, pp 331–339 (2024). https://doi.org/10.1007/978-3-031-54820-8_26
17. Wang, S., Feng, Z.: Mathematical models and exact methods for complex optimization problems. MDPI Mathematics (2023)
18. Hwang, J., Kim, S.: Metaheuristic algorithms for solving real-world engineering optimization challenges. IEEE Access
19. Krishna, G. J. Ravi, V.: Large scale global optimization by hybrid evolutionary computation (2019)

20. Gill, P.E., Wong, E.: Sequential quadratic programming methods. In: Lee, J., Leyffer, S. (eds) Mixed Integer Nonlinear Programming. The IMA Volumes in Mathematics and its Applications, vol 154. Springer, New York, NY (2012). https://doi.org/10.1007/978-1-4614-1927-3_6
21. Plevris, V., & Solorzano, G. (2022). A collection of 30 multidimensional functions for global optimization benchmarking. *Data, 7*(4), 46. https://doi.org/10.3390/data7040046
22. Pan, J.-S., Zhang, Z., Chu, S.-C., Lee, Z.-J., & Li, W. (2023). Application of diversity-maintaining adaptive rafflesia optimization algorithm to engineering optimisation problems. *Symmetry, Algorithms MDPI., 15*(11), 2077. https://doi.org/10.3390/sym15112077
23. Karunarathne, W., Bala, I., Chauhan, D., Roughan, M., Mitchell, L.: Modified CMA-ES algorithm for multi-modal optimization: incorporating niching strategies and dynamic adaptation mechanism. arXiv preprint arXiv:2407.00939 (2024)
24. Farinati, D., & Vanneschi, L. (2024). A survey on dynamic populations in bio-inspired algorithms. *Genet. Program. Evolvable Mach., 25*(2), 19. https://doi.org/10.1007/s10710-024-09492-4
25. Soto, J., Melin, P.: Optimization of the fuzzy integrators in ensembles of ANFIS model for time series prediction: The case of Mackey-Glass. In: Proceedings of the 2015 conference of the international fuzzy systems association and the european society for fuzzy logic and technology (IFSA-EUSFLAT-15), Gijón, Spain, pp 994–999 (2015)
26. Pulido, M., Melin, P., & Castillo, O. (2014). Particle swarm optimization of ensemble neural networks with fuzzy aggregation for time series prediction of the Mexican Stock Exchange. *Information Sciences, 280*, 188–204.
27. Castillo, O., Castro, J. R., Melin, P., & Rodríguez-Díaz, A. (2013). Application of interval type-2 fuzzy neural networks in non-linear identification and time series prediction. *Soft Computing, 18*, 1213–1224.
28. Carreon-Ortiz, H., Valdez, F., Castillo, O.: A new continuous mycorrhiza optimization nature-inspired algorithm. In Springer Briefs in Applied Sciences and Technology. Springer-Verlag GmbH. ISBN: 9783031473685 (2023)
29. Jacobson, S., & Yücesan, E. (2004). Analyzing the performance of generalized hill climbing algorithms. *Journal of Heuristics, 10*, 387–405.
30. Wolpert, D. H., & Macready, W. G. (1997). No free lunch theorems for optimization. *IEEE Transanction on Evolution Computation, 1*(1), 67–82.
31. Geem, Z. W., Kim, J. H., & Loganathan, G. V. (2001). A new heuristic optimization algorithm: Harmony search. *SIMULATION, 76*, 60–68.
32. Nakrani, S., & Tovey, C. (2004). On honey bees and dynamic server allocation in internet hosting centers. *Adaptive Behavior, 12*(3–4), 223–240.
33. Yang, X.S.: A new metaheuristic bat-inspired algorithm. In: González, J.R., Pelta, D.A., Cruz, C., Terrazas, G., Krasnogor, N. (eds) Nature inspired cooperative strategies for optimization (NICSO 2010). Studies in Computational Intelligence, vol 284. Springer, Berlin, Heidelberg (2010). https://doi.org/10.1007/978-3-642-12538-6_6
34. Ramírez, E., Melin, P., & Prado-Arechiga, G. (2019). Hybrid model based on neural networks, type-1 and type-2 fuzzy systems for 2-lead cardiac arrhythmia classification. *Expert. Syst. Appl., 126*, 295–307.
35. Santiago, B., Dorronsoro, A. J., Nebro, J. J., Durillo, O., & Castillo, H. J. (2019). Fraire, A novel multi-objective evolutionary algorithm with fuzzy logic based adaptive selection of operators: FAME. *Information Sciences, 471*, 233–251.
36. Martínez-Soto, R., Castillo, O., & Aguilar, L. T. (2014). Type-1 and Type-2 fuzzy logic controller design using a Hybrid PSO–GA optimization method. *Information Sciences, 285*, 35–49.
37. Castillo, O., Melin, P., Ontiveros, E., Peraza, C., Ochoa, P., Valdez, F., & Soria, J. (2019). A high-speed interval type 2 fuzzy system approach for dynamic parameter adaptation in metaheuristics. *Eng. Appl. Artif. Intell., 85*, 666–680. https://doi.org/10.1016/j.engappai.2019.07.020

Comparative Analysis of the Performance of the Stochastic Fractal Search and Artificial Gorilla Troops Optimization Methods

Marylu L. Lagunes

Abstract The use of optimization algorithms has been growing exponentially in recent years, thanks to their proven efficiency in solving complex problems. This paper describes a comparative analysis of the performance of two nature-based optimization algorithms, the first one Stochastic Fractal Search (SFS) is a method inspired by the natural phenomenon of growth called fractal, this algorithm simulates the form of fractal growth starting with a particle which uses the diffusion property for the creation of random fractals, the second method is the Artificial Gorilla Troop Optimizer (GTO) inspired by the social intelligence of gorilla troops in nature.

Keywords Stochastic fractal search · Artificial gorilla troop optimizer · Optimization algorithms

1 Introduction

Nowadays, optimization is a strategy that is responsible for finding the best solution within a set of possible solutions, which is why it is used in different areas [1, 2, 3 4, 5, 6]. Sometimes simple algorithms are not effective enough to solve problems, and specialized optimization algorithms are needed to provide better performance and solutions to specific problems. There are different classifications for the different optimization methods, some of which we can find are Evolutionary Algorithms [7–12], which are methods based on natural processes such as biological evolution or the collective behavior of organisms, they simulate biological phenomena to find optimal solutions in different search spaces, within the algorithms inspired by nature we find the FSF Algorithm [13] based on the natural growth behavior called fractal which grows exponentially until finding a solution and there is also the GTO method which is inspired by the behavior of gorillas working collectively [14], these two optimization methods have proven to be efficient in solving optimization problems

M. L. Lagunes (✉)
Tijuana Institute of Technology, TecNM, Tijuana, Mexico
e-mail: marylu.lara@tectijuana.mx

O. Castillo and P. Melin (eds.), *Intelligent System Design Based on Soft Computing Models*, Studies in Computational Intelligence 1223,
https://doi.org/10.1007/978-3-031-97309-3_15

in areas such as computing and engineering. Since these two methods have different optimization forms, the objective of this work is to describe a comparative analysis of their performance that can provide valuable conclusions about operation for the solution of different problems.

This paper is organized as follows: Sect. 2 is where we describe the applications of the SFS, in the Sect. 3 the describe the applications of the GTO, in Sect. 4 the comparative analysis of GTO and SFS methods in Sect. 5 the conclusions.

2 Applications of the Stochastic Fractal Search SFS

In this section, the general objective of the optimization problem to be solved with the SFS and GTO methods is briefly explained. These works were taken randomly from recent publications, where the use of the algorithms to be compared is observed for the analysis of their performance in different application areas (Figs. 1, 2).

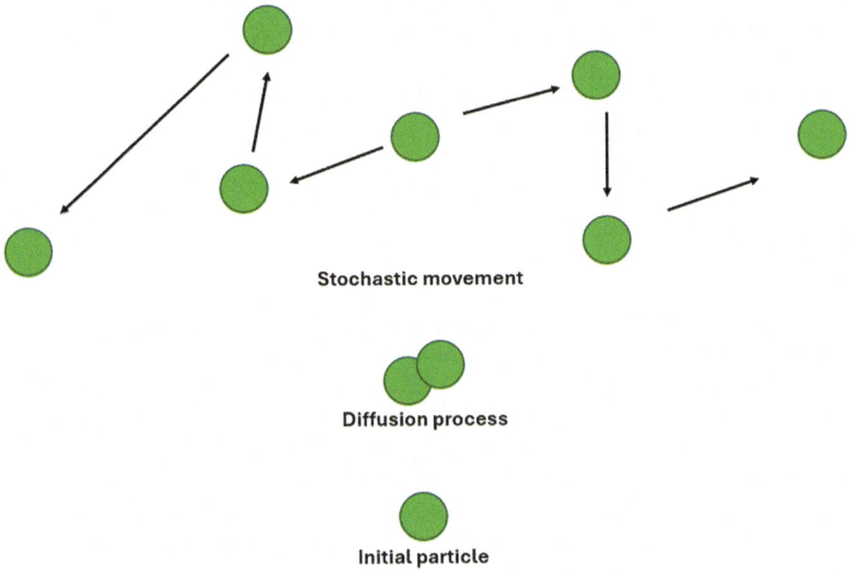

Fig. 1 Representation of fractal movement

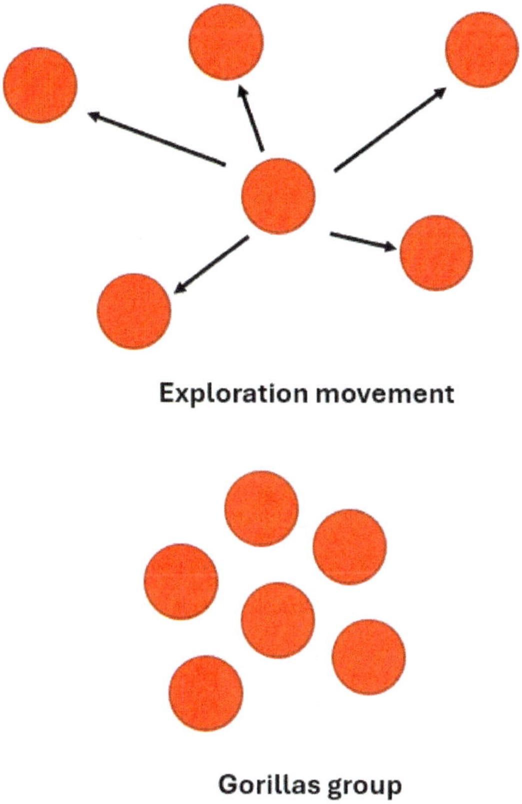

Fig. 2 General movement in GTO

2.1 *Gaussian Barebone Salp Swarm Algorithm with Stochastic Fractal Search for Medical Image Segmentation: A COVID-19 Case Study*

The work in [15] realized a segmentation method based on 2D histogram of nonlocal mean and 2D Kapur entropy called SSA with Gaussian Barebone and Stochastic Fractal Search (GBSFSSSA) by combining the mechanism of Gaussian Barebone and Stochastic Fractal Search. In this work, it is shown that SFS and SSA method combine very well, significantly improving the performance for solving COVID-19 CT image segmentation problems, also GBSFSSSA method was applied to the solution of CEC2017 features demonstrating better efficiency and reliability in solving multi-threshold image segmentation problems.

2.2 A Robust Parameter Estimation Approach Based on Stochastic Fractal Search Optimization Algorithm Applied to Solar PV Parameters

In [16] the optimization of photovoltaic systems is performed, to estimate measured current–voltage (I-V) values. To check the efficiency of the Stochastic Fractal Search method; 3 different case studies are used, the first is the single diode model (SDM), the second is the double diode model (DDM) and third is the photovoltaic module model. The results show that the method used shows a good performance and yield to obtain the parameters of the modeling of solar photovoltaic (PV) cells/modules.

2.3 Stochastic Fractal Search Paradigm in Combination with Artificial Neural Network for Early Prediction of Cooling Load in Residential Buildings

The Stochastic Fractal Search SFS method is used in [17] in conjunction with artificial neural networks (SFS-ANN), for the early prediction of thermal loads, in order to analyze the energy performance in buildings, this hybrid method is used to predict the cooling load (CL) in residential buildings, the results obtained have shown that the use of SFS together with ANN provides a reliable hybrid model for the early prediction of CL, compared to other optimization methods used in prediction experimentation.

3 Applications of Artificial Gorilla Troop Optimizer GTO

This section describes the general objective of the works that use the GTO method as an optimization algorithm.

3.1 Melanoma Diagnosis Using Enhanced Faster Region Convolutional Neural Networks Optimized by Artificial Gorilla Troops Algorithm

The work in [18] shows a novel diagnostic method for skin cancer detection, using the African Gorilla Troops Optimizer (AGTO) method and Faster R-CNN. The model is made according to the experience of doctors and artificial intelligence, the data that is applied to the model is an ISIC-2020 skin cancer dataset. Where the results verify that the performance of the proposed model gives an accuracy of 98.55%.

3.2 Automatic Tuning of the PID Controller Based on the Artificial Gorilla Troops Optimizer

Currently heuristics are used to do the tuning of PID controllers, in this work [19] the PID tuning is proposed using the artificial gorilla troop optimizer (GTO) algorithm, creating a hybrid adaptive PID_GTO predictive model based on the artificial gorilla troop optimizer (GTO) algorithm. The errors used to measure the performance of the model are the integral time squared error (ITSE), absolute integral time error (ITAE), absolute integral error (IAE) and integral squared error (ISE), the results show that it outperforms previous techniques and gives stability to the DC motor.

3.3 UAV Cluster Task Assignment Algorithm Based on Improved Artificial Gorilla Troops Optimize

In this work [20] an improved version of Artificial Gorilla Troop Optimizer (GTO) algorithm is proposed, adopting Halton sequence to generate the initial population to ensure diversity, to demonstrate its effectiveness it is used to optimize a group of Unmanned Aerial Vehicles (UAV) working collaboratively by assigning various tasks to each UAV effectively. It is based on the search for information exchange to improve the communication between the gorilla and the population, effectively jumping from the local optimal solution, the results showed that the GTO algorithm outperformed other previously used methods giving fast and stable assignments.

4 Comparative Analysis of GTO and SFS Methods

This section describes the comparison tables with the main characteristics of the methods used, in order to better understand how each of the algorithms is structured.

As shown in Table 1, the SFS method is controlled by the diffusion and particle creation property, which tells us that this algorithm has a good performance in complex problems due to its stochastic exploration where the search space is wide.

The GTO algorithm has a social behavior imitating the natural behavior of gorillas, they are based on the collaboration of food search, exploring and exploiting the search space.

These two methods depend on the size of their population, which is the number of particles in SFS and the number of gorillas in GTO. The first is based on the natural fractal creation, which explores and exploits as its size increases and tends to have a fast but effective convergence. The second is based on its social behavior of imitation and collaboration, which makes them explore and exploit the search space. Its convergence is slow but balanced and therefore generates good performance in problem solving.

Table 1 Main features of SFS and GTO algorithms

Factor	Stochastic fractal search SFS	Artificial gorilla troops optimizer
Inspiration	Inspired by the natural phenomenon of stochastic fractal growth	Inspired by the cooperative social behavior of gorillas when foraging
Behavior structure	Particle solutions are explored using fractal inspiration and stochastic strategy	Population of solutions that mimic the cooperative interactions and strategies of gorilla troops
Key parameters	Number of fractals, dispersion rate, exploration and exploitation coefficient	Population number, exploration and exploitation factors, convergence coefficient
Main operators	Stochastic generation of fractal solutions and updating by scattering and exploitation	Gorilla behavior based on imitation and leadership, balancing exploration and exploitation
Speed of convergence	Fast convergence in the early stages, especially when using good fractal dispersion	Moderate convergence, stability across iterations, and balanced between exploration and exploitation

Table 2 shows comparison of method usage, where the first column gives the references of the analyzed article, the second indicates whether the method was used originally, the third column shows whether the algorithm was hybridized and the last tells us whether a good performance of the method was obtained in solving the given problem.

Table 2 shows comparison of method usage, where the first column gives the references of the analyzed article, the second indicates whether the method was used originally, the third column shows whether the algorithm was hybridized and the last tells us whether a good performance of the method was obtained in solving the given problem.

Table 2 Comparative table of the use of methods

Reference	Original	Hybrid	Better performance
1	No	Yes	Yes
2	No	Yes	Yes
3	Yes	No	Yes
4	No	Yes	Yes
5	No	Yes	Yes
6	Yes	No	Yes

5 Conclusions

In the comparative study, recent works were analyzed [21], three from GTO and three from SFS from different areas and optimization problems.

It can be concluded that, with the comparison of the methods, we learned that the SFS algorithm has a fast convergence and a good exploration, it usually takes time to generate optimal solutions due to the stochastic dispersion of the particles, however, it has a good performance for solutions in complex problems, and the GTO method converges slowly due to its capacity for social exploitation, it has good performance in continuous optimization problems. In addition, they proved to be compatible with other methods, creating efficient hybridizations for the solution of robust problems. In future work, we could try to apply these algorithms in other problems [22], or compare with other metaheuristics [23–25].

References

1. Połap, D., & Woźniak, M. (2021). Red fox optimization algorithm. *Expert. Syst. Appl., 166*, Article 114107.
2. Faramarzi, M., Heidarinejad, S.M., Gandomi A.H.: Marine predators algorithm: A nature-inspired metaheuristic. Expert. Syst. Appl 152, 113377 (2020)
3. Schlipf, M., & Gygi, F. (2015). Optimization algorithm for the generation of ONCV pseudopotentials. *Comput. Phys. Commun., 196*, 36–44.
4. Zhong, C., Li, G., & Meng, Z. (2022). Beluga whale optimization: A novel nature-inspired metaheuristic algorithm. *Knowl.-Based Syst., 251*, Article 109215.
5. Hu, G., Zheng, Y., Abualigah, L., & Hussien, A. G. (2023). DETDO: An adaptive hybrid dandelion optimizer for engineering optimization. *Adv. Eng. Inform., 57*, Article 102004.
6. Khishe, M., Mosavi, M.R.: Chimp optimization algorithm. Expert. Syst. Appl. 149, 113338 (2020)
7. Holland, J. H. (1992). Genetic algorithms. *Scientific American, 267*(1), 66–73.
8. Bäck, T. (1995). Evolution strategies: An alternative evolutionary algorithm. *In Eur. Conf. Artif. Evol., 1063*, 1–20.
9. Qin, K., Huang, V. L., & Suganthan, P. N. (2008). Differential evolution algorithm with strategy adaptation for global numerical optimization. *IEEE Trans. Evol. Comput., 13*(2), 398–417.
10. Kaveh, T.B., Afshari, E.: An efficient hybrid particle swarm and swallow swarm optimization algorithm. Comput. Struct. 143, 40–59 (2014)
11. Marco, D.: Ant colony optimization and swarm intelligence. In: 6th international conference, ANTS 2008, Brussels, Belgium, September 22–24, 2008, Proceedings. Springer (2008)
12. Xu, Y., Fan, P., Yuan, L.: A simple and efficient artificial bee colony algorithm. Math. Probl. Eng 2013, 526315 (2013)
13. H. S.-K.-B. Stochastic fractal search: a powerful metaheuristic algorithm. Knowl.-Based Syst. 75, 1–18 (2015)
14. Abdollahzadeh, F., Soleimanian, G., Mirjalili, S.: Artificial gorilla troops optimizer: A new nature-inspired metaheuristic algorithm for global optimization problems. Int. J. Intell. Syst 36(10), 5887–5958 (2021)
15. Zhang, Q., et al. (2021). Gaussian barebone salp swarm algorithm with stochastic fractal search for medical image segmentation: A COVID-19 case study. *Computers in biology and medicine, 139*, 10494.

16. Rezk, H., Babu, T., Al-Dhaifallah, M., H. Z.-E.: A robust parameter estimation approach based on stochastic fractal search optimization algorithm applied to solar PV parameters. Energy Reports 7, 620–640 (2021)
17. Moayedi, H.A. M.: Suggesting a stochastic fractal search paradigm in combination with artificial neural network for early prediction of cooling load in residential buildings. Energies 14(6), 1649 (2021)
18. Nivedha, S. S.: Melanoma diagnosis using enhanced faster region convolutional neural networks optimized by artificial gorilla troops algorithm. Inf. Technol. Control. 52(4), 819 (2023)
19. Abbas, M.K.-I.J. (2024). Automatic tuning of the PID controller based on the artificial gorilla troops optimizer. *Iraqi J. Sci., 65*(5), 2867–2880.
20. Zhang, R., Ren, H., Li, X., D.I, Y.: UAV Cluster task assignment algorithm based on improved artificial gorilla troops optimizer. IEEE Access 11, 135133–135146, (2023)
21. Lagunes, M. L., Miramontes, I., Castillo, O., & Melin, P. (2024). Comparative study of gorilla troops optimizer and stochastic fractal search with fuzzy dynamic parameter adaptation. *Computación y Sistemas, 28*(2), 2024.
22. Ramírez, E., Melin, P., & Prado-Arechiga, G. (2019). Hybrid model based on neural networks, type-1 and type-2 fuzzy systems for 2-lead cardiac arrhythmia classification. *Expert. Syst. Appl., 126*, 295–307.
23. Santiago, B., Dorronsoro, A. J., Nebro, J. J., Durillo, O., & Castillo, H. J. (2019). Fraire, A novel multi-objective evolutionary algorithm with fuzzy logic based adaptive selection of operators: FAME. *Information Sciences, 471*, 233–251.
24. Martínez-Soto, R., Castillo, O., & Aguilar, L. T. (2014). Type-1 and Type-2 fuzzy logic controller design using a Hybrid PSO–GA optimization method. *Information Sciences, 285*, 35–49.
25. Castillo, O., Melin, P., Ontiveros, E., Peraza, C., Ochoa, P., Valdez, F., & Soria, J. (2019). A high-speed interval type 2 fuzzy system approach for dynamic parameter adaptation in metaheuristics. *Eng. Appl. Artif. Intell., 85*, 666–680. https://doi.org/10.1016/j.engappai.2019.07.020

Genetic Optimization of PID Controllers for Unmanned Aerial Vehicles

Cesar Garcia, Fevrier Valdez, and Oscar Castillo

Abstract In this paper we are presenting the genetic optimization of the gains in proportional-integral-derivative (PID) controllers for unmanned aerial vehicles. Genetic algorithms are applied to find the optimal values of the gains in the PID controllers so the drone can follow a given desired trajectory. The results of the optimized controller are compared against a non-optimized PID controller to show the advantages of utilizing genetic algorithms for this task.

Keywords Genetic algorithm · Control · Optimization

1 Introduction

Evolutionary algorithms, inspired by the evolutionary behavior of species, aim to solve optimization and search problems with high efficiency. The key to their success lies in tuning and controlling their parameters, allowing them to find high-quality solutions using fewer iterations and less computational effort [1–4].

One of the main algorithms developed from this observation is the genetic algorithm (GA). This algorithm is an optimization technique based on the behavior of groups of individuals during evolution, which enables the exploration of a wide range of potential solutions in line with the objective function being optimized. By simulating the evolutionary process, GA leverages the competition of the group to iteratively improve upon solutions, making it highly effective for complex optimization problems [5], like the one being considered in this paper. Of course, there are other optimization algorithms that could be considered in the future, such as the ones in [6–10].

The development of autonomous flight systems has grown lately because of the increasing amount of applications of unmanned aerial vehicle (UAV) in military fields, in civil fields and many others that are emerging [11]. In modern times, there

C. Garcia · F. Valdez (✉) · O. Castillo
Tijuana Institute of Technology, TecNM, Tomas Aquino, Tijuana, B.C, México
e-mail: fevrier@tectijuana.mx

O. Castillo and P. Melin (eds.), *Intelligent System Design Based on Soft Computing Models*, Studies in Computational Intelligence 1223,
https://doi.org/10.1007/978-3-031-97309-3_16

is a wide variety of aircrafts used for UAVs, like fixed-wing airplanes, airships, and helicopters, among others [12]. Despite the advantages of quadcopters, like other aircrafts, they are very unstable systems, and so any imbalance in their motion (in particular for roll and pitch) generates angular and linear accelerations, which can produce a collision if not compensated in a quick fashion. Moreover, the quadcopter system is nonlinear and the movements are coupled, which causes even more complexity. PID controllers offer a simple, but at the same time an effective solution to stabilize the aircraft because they make it possible to treat every variable independently within a limited range in which the behavior of the quadcopter is approximately linear [13, 14].

This article proposes the utilization of genetic algorithms for finding the optimal gains of the PID controllers for the drone, so that it can follow a desired trajectory and avoid obstacles. This approach is able to improve the performance of non-optimized PID controllers showing the advantages of the proposal.

The rest of the paper is organized as follows. In Sect. 2 the model of drone motion is presented. Section 3 outlines the PID control method. In Sect. 4 the results are summarized. Section 5 offers the conclusions of this work.

2 Model of Drone Motion

In this section we briefly show the elementary concepts involved in the motion of a drone [11, 12]. In Fig. 1 we can appreciate the three main orientation angles for a drone: pitch, roll and yaw.

The model of the drone includes the three main orientation angles: pitch, roll and yaw. The model consists of a set of nonlinear differential equations, but for simplicity we use a linearized version of the same, as follows.

$$\ddot{\theta} = \frac{1}{I_\theta}\left(T_\theta - D_\theta\dot{\theta}\right) \tag{1}$$

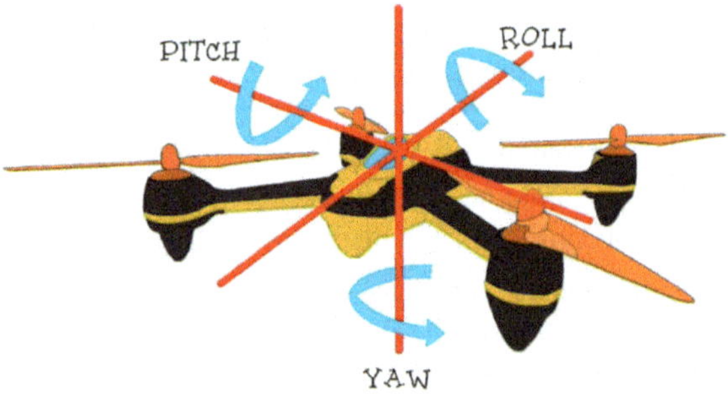

Fig. 1 Illustration of the three main orientation angles in the drone motion

$$\ddot{\phi} = \frac{1}{I_\emptyset}\left(T_\phi - D_{\dot{\phi}}\,\dot{\Phi}\right) \quad (2)$$

$$\ddot{\psi} = \frac{1}{I_\psi}\left(T_\psi - D_{\dot{\psi}}\,\dot{\psi}\right) \quad (3)$$

where:

- $\ddot{\theta}$, $\ddot{\phi}$, $\ddot{\psi}$ represent the Pitch, Roll, and Yaw, respectively.
- $I_{\emptyset,} I_\emptyset I_\psi$ are the inertia moments around the axis.
- $T_{\emptyset,} T_\emptyset T_\psi$ are the applied torques.
- $D_{\emptyset,} D_\emptyset D_\psi$ are the damping coefficients.

We can utilize this model to simulate the behavior of the drone and test the proposed genetic optimization of the gains used in the PID controller.

3 PID Control Method

In this section we define the utilized PID controller of the drone. Equation (4) defines the general PID control scheme, which consists of three parts: the error, the derivative of error and the integral of the error.

$$u(t) = K_p e(t) + K_i \int e(t)dt + K_d \frac{de(t)}{dt} \quad (4)$$

where:

- u(t) is the control signal.
- e(t) is the error (difference between the desired value and the real value).
- K_p is the proportional gain constant.
- K_i is the integral gain constant.
- K_d is the derivative gain constant.

These gains are the ones that are found by the genetic algorithm approach, so that the PID controller behaves in the best way possible for following the desired trajectories.

4 Results

Now we are presenting the results of the genetic application to find the optimal gains of the PID fuzzy controllers for pitch, roll and yaw. Table 1 summarizes the gains (Kd, Ki and Kp) obtained for pitch, roll and yaw for 30 experiments that were performed with the genetic algorithm.

Table 1 Gains obtained by the genetic algorithm for pitch, roll and yaw

EX	GEN	Kd pitch	Kd roll	Kd yaw	Ki pitch	Kiroll	Ki yaw	Kp pitch	Kp roll	Kp yaw
1	10	0.2347	0.6392	0.8101	8.1	9.5	0	9.8	9.8	9.9
2	20	0.0149	0.6610	0.0992	9.9	9.1	10	9.9	9.7	9.9
3	30	0.0105	0.6657	0.0080	10.0	10.0	10	10.0	10.0	10.0
4	40	0.0025	0.6646	0.0019	10.0	10.0	10	10.0	10.0	10.0
5	50	0.0062	0.6643	0.0002	10.0	10.0	10	10.0	10.0	10.0
6	60	0.0098	0.6635	0.0015	10.0	9.9	10	10.0	10.0	10.0
7	70	3E-05	0.6620	4E-05	10.0	10.0	10	10.0	10.0	10.0
8	80	0	0.6636	0.9445	10.0	10.0	0	10.0	10.0	10.0
9	90	7E-05	0.6636	0	10.0	10.0	10	10.0	10.0	10.0
10	100	0.0002	0.6633	8E-05	10.0	10.0	10	10.0	10.0	10.0
11	120	3E-05	0.6747	0.0001	10.0	10.0	10	10.0	10.0	10.0
12	130	3E-05	0.6747	1E-05	10.0	10.0	10	10.0	10.0	10.0
13	140	0.0001	0.6748	0.9449	10.0	10.0	0	10.0	10.0	10.0
14	150	3E-06	0.6633	0	10.0	10.0	10	10.0	10.0	10.0
15	160	2E-05	0.6748	0.9446	10.0	10.0	0	10.0	10.0	10.0
16	170	1E-05	0.6747	3E-05	10.0	10.0	10	10.0	10.0	10.0
17	180	0.0001	0.6751	0.9447	10.0	10.0	0	10.0	10.0	10.0
18	190	4E-05	0.6749	0	10.0	10.0	10	10.00	10.0	10.0
19	200	9E-07	0.6633	0.9448	10.0	10.0	0	10.00	10.0	10.0
20	210	2.E-05	0.6747	2E-05	10.0	10.0	10	10.00	10.0	10.0
21	220	0.0001	0.6751	5E-05	10.0	10.0	10	10.00	10.0	10.0
22	230	4E-05	0.6749	3E-05	10.0	10.0	10	10.00	10.0	10.0
23	240	8E-05	0.67483	4E-06	10.0	10.0	10	10.00	10.0	10.0
24	250	0.0001	0.6748	8E-05	10.0	10.0	10	10.00	10.0	10.0
25	260	2E-05	0.6747	0	10.0	10.0	10	10.00	10.0	10.0
26	300	8E-05	0.6633	0.9448	10.0	10.0	0	10.00	10.0	10.0
27	320	0	0.6748	0.0005	10.0	10.0	10	10.00	10.0	10.0
28	350	4E-07	0.6634	0.9448	10.0	10.0	0	10.00	10.0	10.0
29	400	1E-06	0.6637	4E-05	10.0	10.0	10	10.0	10.0	10.0
30	450	7E-05	0.6632	0.9448	10.0	10.0	0	10.0	10.0	10.0

We show in Figs. 2 and 3 two different experiments with the drone following specific trajectories. Figure 2 shows a simulation that produces mean squared errors (MSEs) of 0.000068256 for the Pitch, 0.0033999 for the Roll, and 0.00035137 for the Yaw, respectively. Figure 3 shows a simulation that produces mean squared errors (MSEs) of 0.11014 for the Pitch, 0.026608 for the Roll, and 0.62518 for the Yaw, respectively. As can be noted the second case (of Fig. 3) has higher errors, but still

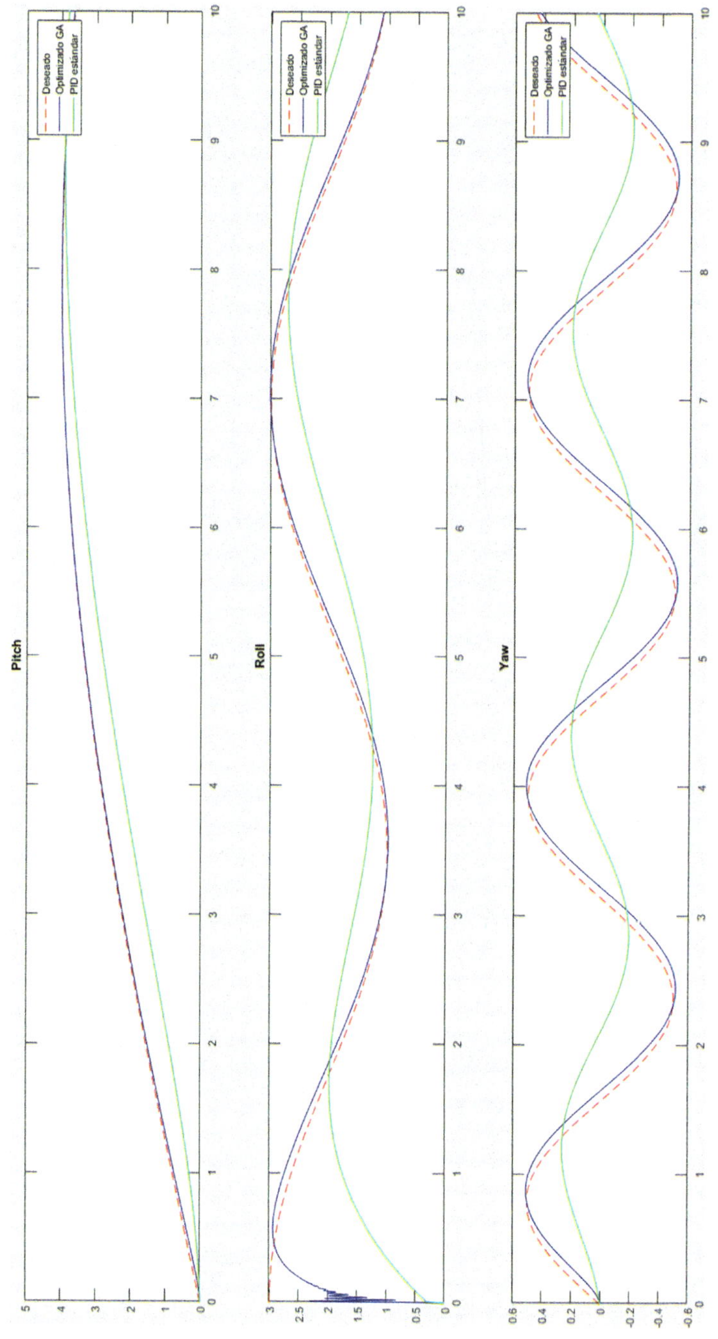

Fig. 2 Illustration of the proposal for experiment 1

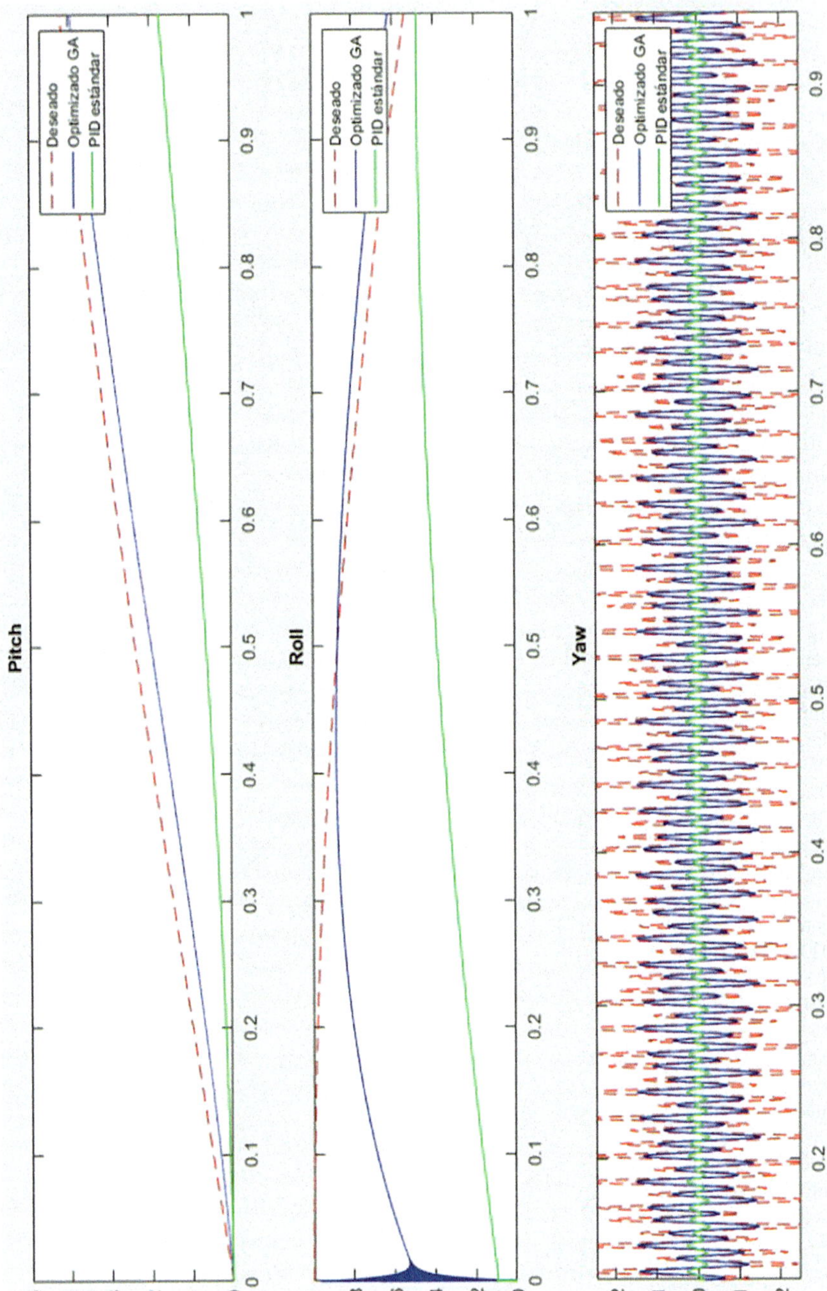

Fig. 3 Illustration of the proposal for experiment 2

the proposed optimized controller outperforms the PID controller. In the figures, we have the blue signal for the optimized controller, the green signal as the one provided by the PID controller and the dotted red signal as the desired trajectory.

5 Conclusions

In this paper we have described the genetic optimization of the gains in proportional-integral-derivative (PID) controllers for unmanned aerial vehicles. Genetic algorithms were applied to find the optimal values of the gain parameters in the PID controllers so the drone can follow a given desired trajectory. The results of the optimized controller were compared against a non-optimized PID controller to show the advantages of utilizing genetic algorithms for this task. As future work, we plan to consider fuzzy PID controllers and their optimization with the idea of achieving even better results. We can also consider other metaheuristics for performing the optimization [15–21].

References

1. Sennan, S., Ramasubbareddy, S., Balasubramaniyam, S., Nayyar, A., Abouhawwash, M., & Hikal, N. A. (2021). T2FL-PSO: type-2 fuzzy logic-based particle swarm optimization algorithm used to maximize the lifetime of internet of things. *IEEE Access, 9*, 63966–63979. https://doi.org/10.1109/ACCESS.2021.3069455
2. Havens, T., Spain, C., Salmon, N., & Keller, J. (2008). Roach infestation optimization. In *2008 IEEE Swarm Intelligence Symposium*. https://doi.org/10.1109/sis.2008.4668317.
3. Olivas, F., Valdez, F., Castillo, O., & Melin, P. (2016). Dynamic parameter adaptation in particle swarm optimization using interval type-2 fuzzy logic. *Soft Computing, 20*(3), 1057–1070. https://doi.org/10.1007/s00500-014-1567-3
4. Sharmila, S., & Vijayarani, S. (2021). Association rule mining using fuzzy logic and whale optimization algorithm. *Soft Computing, 25*(2), 1431–1446. https://doi.org/10.1007/s00500-020-05229-4
5. Rodríguez-Abreo, O., Rodríguez-Reséndiz, J., García-Cerezo, A., García-Martínez, J.R. (2024). Fuzzy logic controller for UAV with gains optimized via genetic algorithm. *Heliyon, 10*(4)
6. Amador-Angulo, L., & Castillo, O. (2023). An Interval Type-2 fuzzy logic approach for dynamic parameter adaptation in a whale optimization algorithm applied to mathematical functions. *Axioms, 13*(1), 33. https://doi.org/10.3390/axioms13010033
7. Obagbuwa, I., Adewumi, A., & Adebiyi, A. (2014). A dynamic step-size adaptation roach infestation optimization. In *2014 IEEE International Advance Computing Conference (IACC)*. https://doi.org/10.1109/iadcc.2014.6779498.
8. Lagunes, M., Miramontes, I., Castillo, O., & Melin, P. (2024). Comparative study of gorilla troops optimizer and stochastic fractal search with fuzzy dynamic parameter adaptation. *Computación y Sistemas, 28*(2). https://doi.org/10.13053/cys-28-2-5028.

9. Nava, R., Kavitha, M., Reddy, G., Yousef, A., AboRas, K., Emara, A., & Reddy, C. (2024). Mayfly optimistic hyperelliptic curve cryptosystem. *Frontiers in Computer Science, 6*. https://doi.org/10.3389/fcomp.2024.1381850.
10. Rajabioun, R. (2011). Cuckoo optimization algorithm. *Applied Soft Computing, 11*(8), 5508–5518. https://doi.org/10.1016/j.asoc.2011.05.008
11. Song, Z., & Huang, H. (2024). Safety-Critical fixed-time formation control of quadrotor UAVs with disturbance based on robust control barrier functions. *Drones, 8*, 618. https://doi.org/10.3390/drones8110618
12. Kuang, J., & Chen, M. (2024). Adaptive sliding mode control for trajectory tracking of quadrotor unmanned aerial vehicles under input saturation and disturbances. *Drones, 8*, 614. https://doi.org/10.3390/drones8110614
13. Li, Y., Wen, D., Zhang, S., & Li, L. (2024). Sequential task allocation of more scalable artificial dragonfly swarms considering dubins trajectory. *Drones, 8*, 596. https://doi.org/10.3390/drones8100596
14. Sui, Y., Yang, Z., Zhuo, H., You, Y., Que, W., & He, N. (2024). A fuzzy pure pursuit for autonomous ugvs based on model predictive control and whole-body motion control. *Drones, 8*, 554. https://doi.org/10.3390/drones8100554
15. Carreon-Ortiz, H., Valdez, F., & Castillo, O. (2023). Comparative study of type-1 and interval type-2 fuzzy logic systems in parameter adaptation for the fuzzy discrete mycorrhiza optimization algorithm. *Mathematics, 11*(11). https://doi.org/10.3390/math11112501.
16. Bernal, E., Lagunes, M., Castillo, O., Soria, J., & Valdez, F. (2021). Optimization of type-2 fuzzy logic controller design using the GSO and FA algorithms. *International Journal of Fuzzy Systems, 23*(1), 42–57. https://doi.org/10.1007/s40815-020-00976-w
17. Castillo, O., Melin, P., Ontiveros, E., Peraza, C., Ochoa, P., Valdez, F., & Soria, J. (2019). A high-speed interval type 2 fuzzy system approach for dynamic parameter adaptation in metaheuristics. *Engineering Applications of Artificial Intelligence, 85*, 666–680. https://doi.org/10.1016/j.engappai.2019.07.020
18. Santiago, B., Dorronsoro, A. J., Nebro, J. J., Durillo, O., & Castillo, H. J. (2019). Fraire, A novel multi-objective evolutionary algorithm with fuzzy logic based adaptive selection of operators: FAME. *Information Sciences, 471*, 233–251.
19. Martínez-Soto, R., Castillo, O., & Aguilar, L. T. (2014). Type-1 and Type-2 fuzzy logic controller design using a Hybrid PSO–GA optimization method. *Information Sciences, 285*, 35–49.
20. Amador-Angulo, L., Castillo, O., Melin, P., Geem, Z.G. (2024). Type-3 fuzzy dynamic adaptation of Bee colony optimization applied to mathematical functions. *Fuzzy Sets and Systems, 489*. https://doi.org/10.1016/j.fss.2024.109014
21. Ochoa, P., Peraza, C., Melin, P., Castillo, O., Park, S., & Geem, Z. W. (2024). Enhancing control systems through type-3 fuzzy logic optimization. *Mathematics, 12*, 1792. https://doi.org/10.3390/math12121792

The manufacturer's authorised representative in the EU is Springer
Nature Customer Service Centre GmbH, Europaplatz 3, 69115 Heidelberg,
Germany. If you have any concerns regarding our products, please
contact ProductSafety@springernature.com

Printed and bound by CPI Group (UK) Ltd, Croydon, CR0 4YY
01/10/2026
02225883-0018